Statistics Laboratory Manual
Experiments Using Minitab

Gerald Keller
Wilfrid Laurier University

 Duxbury Press
An Imprint of Wadsworth Publishing Company
Belmont, California

Statistics Editor: Curt Hinrichs
Editorial Assistant: Jenny Burger
Production Editor: Carol Carreon Lombardi
Copy Editor: Barbara Kimmel
Print Buyer: Diana Spence
Compositor: Laurel Tutoring
Printer and Binder: Malloy Lithographing
Cover Design: Stuart Paterson

*This book is printed on
acid-free recycled paper.*

I⟨T⟩P ™

International Thomson Publishing
The trademark ITP is used under license

Duxbury Press
An Imprint of Wadsworth Publishing Company
A division of Wadsworth, Inc.

Printed in the United States of America

ISBN 0-534-21864-4

1 2 3 4 5 6 7 8 9 10 — 98 97 96 95 94

CONTENTS

PREFACE

For students of statistics to become practitioners of statistics, they must develop several skills. The first is the ability to employ the correct statistical technique to solve a specific problem. This ability, together with a computer and a software package, is sufficient to produce results. To interpret these results requires a second skill: an understanding of the conceptual foundation upon which statistics is based. It is possible to gain this understanding by reading textbooks or attending lectures. However, we believe that an understanding of statistical principles is achieved more easily by active experimentation. This manual describes a number of experiments that allow students to answer a number of "What happens if . . . " questions. For example, students are taught that a 95% confidence interval estimate is interpreted to indicate that, in repeated sampling, 95% of the confidence interval estimates produced would contain the true value of the parameter being estimated. In Chapter 4 we present experiments that demonstrate precisely what this means. We teach that the statistic $(\bar{x} - \mu)/(s/\sqrt{n})$ is Student t distributed provided that the population from which the sample was drawn is normally distributed. In this manual we provide several experiments that allow students to discover what happens when the normality requirement is not satisfied.

All experiments use the Minitab statistical software package. We assume that users of this manual have access to a computer with Minitab software and know how to access the system. We further assume that students are currently taking a statistics course and that each topic covered by the experiments will have been presented to the students either through textbook reading or class discussions. By observing the results of the experiments, students will gain a much deeper comprehension of the material than is possible through a passive approach.

We have tested each of the experiments ourselves and found the outcomes consistent with the expected theoretical results. If there are any problems, we would appreciate being notified. We would also like to hear from students or instructors with suggestions about other experiments or criticism about this manual. Write to Dr. Gerald Keller, c/o Wadsworth Publishing Company, 10 Davis Drive, Belmont, California 94002.

We would like to thank the following colleagues for their help and suggestions during the development of this manuscript: Jay Devore, California Polytechnic State University, San Luis Obispo; Melvin Ott, Medical Service Bureau; David M. Rocke, University of California, Davis; Marlene A. Smith, University of Colorado at Denver; and Bruce E. Trumbo, California State University, Hayward.

CHAPTER 1

Learning Statistical Concepts

1.1 INTRODUCTION

Anyone who plans to use statistical techniques must have an understanding of the concepts and principles that underlie the statistical procedures. Most students attempt to learn these principles by reading a textbook or by passively sitting through lectures. This manual was written to provide an alternative way of understanding statistical concepts. We believe that most theoretical ideas are comprehended more easily by active experimentation. Just as in chemistry or physics, performing experiments, observing results, and reporting conclusions provide a deeper insight about the subject than is possible by only reading books or listening to professors. This manual describes a variety of experiments students of statistics can perform using the Minitab software system that will allow them to make "discoveries" about the underlying principles of statistics.

There are differences between the kind of experiment one performs in chemistry or physics and the kind one performs in statistics. In a chemistry or physics experiment, if we repeat the procedures precisely we will observe exactly the same result (except for relatively small measurement errors). For example, dropping an object in a vacuum and measuring its speed after one second will produce the same speed no matter how frequently we do it (assuming that the location of the experiment is unchanged). However, the very nature of probability and statistics makes identical results from the same experiment almost impossible. For example, if our experiment consists of flipping a fair coin 100 times and counting the number of heads, the results will probably be somewhere between 40 and 60 heads. If we perform the experiment twice, seeing the same number of heads in both experiments is quite unlikely. Because of the randomness that is part of the type of experiments that we conduct in statistics, it is necessary for us to repeat that experiment a number of times in order to draw the correct conclusion. Thus, if we flip two fair coins 100 times and observe that two heads occurred 30 times, we may erroneously conclude that the probability of observing two heads in flips of two fair coins is 30%. However, if we repeat the experiment 10,000 times and count the number of times we observed two heads, we'd see a number quite close to 2500. Performing the experiment a large number of times would allow us to confirm that the probability of seeing two heads in flips of two fair coins is 25%.

Just as in a chemistry course, you will be expected to summarize what you have learned from each experiment. To help you do this, we have supplied you with an outline to report your findings. These outlines will emphasize the critical principles that you're expected to grasp as a result of performing the experiments.

In each of the following eight chapters we describe experiments that will provide insights into some aspect of statistics. Each chapter begins with a brief introduction of the concepts involved. However, these introductions are not designed to teach you all about the concepts. Instead they're only meant to review the critical principles. You are expected to have covered this material either in a statistics course you are now taking or in one recently completed. Ideally, you should have some understanding of the material so that you can benefit fully from the experiments.

1.2   USING MINITAB

All the experiments will be conducted using the Minitab random number generator. The command[*]

> *RANDOM 100 C1*

causes Minitab to create 100 random numbers and store them in column 1. We can specify the distribution from which the numbers are to be drawn. For example, the following command and subcommand

> *RANDOM 100 C1;*

> *INTEGER 0 9.*

will randomly draw 100 integers between 0 and 9 (with each digit equally likely) and put them in column 1. The results can be seen by issuing the command

> *PRINT C1*

The following output was generated when we tried this experiment.

> *2 3 9 6 9 8 4 3 1 7 6 1 5 1 9 9 1 1 1 6 2 5 3 9 8 8 1 8 6 5*

> *0 8 1 8 4 9 3 8 4 7 6 9 3 1 2 0 7 6 3 5 4 8 8 1 5 9 0 3 9 4*

> *3 8 0 2 7 3 5 1 4 8 4 2 6 0 5 0 6 2 8 1 7 1 3 2 8 3 3 3 9 4*

> *8 5 6 3 8 0 7 9 8 9*

It may be helpful for you to picture a container holding a very large (in fact, infinite) number of slips of paper, each containing an integer between 0 and 9. If we could count the number of slips of paper containing each integer, we would see that the integers occur with exactly the same frequency—10%.

---

[*]Minitab reads only the first four letters of any command or subcommand. Thus, for example, we can type *RAND* instead of *RANDOM*. Nevertheless, we will always use the complete command name in this manual to ensure clarity. Be sure to input semicolons and periods as shown.

Throughout this manual we will provide a variety of subcommands to generate observations from different distributions. Thus, the values of the random variables and the frequencies with which they occur will be specified. We will also give you instructions on how to calculate the relevant statistics and output the results. Your job will be to interpret the outcomes to confirm your comprehension of the theory. The following example illustrates this point.

EXAMPLE 1.1

We can simulate flipping two fair coins 100 times in the following way:

RANDOM 100 C1 C2;

BERNOULLI .5.

The command tells Minitab to generate 100 random numbers in each of columns 1 and 2. The subcommand further instructs Minitab that the numbers are to be generated from a Bernoulli process with $p = .5$. If you were to issue the command

PRINT C1 C2

you would see in each column a series of 0s and 1s. A Bernoulli process is the process that produces a binomial distribution. Thus, the 1s represent successes and the 0s represent failures. For our purposes the 1s can represent heads and the 0s can indicate tails. Each row can be interpreted to represent the results of conducting the experiment once, where the experiment consists of flipping two fair coins. To count the number of events representing the flips of the two coins, type

TABLE C1 C2

When we tried this example, the printout below was produced.

ROWS: C1 COLUMNS: C2

	0	*1*	*ALL*
0	21	28	49
1	21	30	51
ALL	42	58	100

CELL CONTENTS—COUNT

Using the notation described above, the printout tells us that the experiment yielded 21 outcomes where both flips resulted in tails (0,0), 28 outcomes where the first flip resulted in tails and the second flip resulted in heads (0,1), 21 outcomes where the first flip was heads and the second flip was tails $(1,0)$, and 30 outcomes where both flips resulted in heads $(1,1)$. We can summarize all the output as follows:

Event	Frequency	Relative Frequency
TT $(0,0)$	21	.21
TH $(0,1)$	28	.28
HT $(1,0)$	21	.21
HH $(1,1)$	30	.30

We should not conclude that the relative frequencies produced by this experiment yield the correct probabilities associated with flipping two fair coins. That is, we should not conclude from our experiment that $P(HH) = .30$. To estimate the probability of observing two heads in flips of two fair coins, we should repeat the experiment a large number of times. Even then it is quite unlikely that the relative frequency of two heads would equal exactly 25%. However, the larger the number of times the experiment is repeated, the closer the relative frequency will be to the theoretical probability of 25%. Using the normal approximation to the binomial distribution, we can determine that, if the experiment is conducted 3000 times, the relative frequency of two heads would almost certainly (99% of the time) lie within 2% of the true value of 25%. Performing the experiment 12,000 times would likely yield a relative frequency that falls within 1% of the true value. This point raises an issue that cannot be further postponed; that is, how many times can we repeat the experiment? The answer depends on the amount of worksheet space available in the version of Minitab that you're using.

Most mainframe computer versions of Minitab have worksheet capacities of 50,000 entries. The most recent full (not student) PC versions have the capability of storing about 16,000 entries. The latest student versions are capable of storing 3500 numbers. In the experiments that we describe in this manual, we will assume that you have a full PC version and a worksheet capacity of at least 16,000 entries. However, to assist those who only have access to a student version, we provide separate instructions that reduce the size of the experiment so that a worksheet of 3500 entries will suffice.

1.3 MINITAB INSTRUCTIONS

We assume that you are familiar with how to use Minitab; however, you do not have to be an expert. In this section we outline some of the critical instructions that you will need in this manual.

Generating Random Numbers

The command *RANDOM K C1* creates K random numbers and stores them in column 1. Subcommands can be used to specify the distribution of the random numbers and relevant parameters. The subcommands are listed below.

INTEGER K1 K2 specifies that the random numbers are integers between K1 and K2 with each integer equally likely.

NORMAL K1 K2 tells Minitab that you want the values to be drawn from a normal distribution with mean = K1 and standard deviation = K2.

BERNOULLI K1 instructs Minitab to generate a series of 0s and 1s from a Bernoulli process where the probability of observing a 1 is K1.

EXPONENTIAL K1 specifies that the distribution is exponential with mean = K1.

UNIFORM K1 K2 tells Minitab that the distribution is uniform between K1 and K2. That means that the values generated can lie anywhere between K1 and K2.

Don't forget to end each command with a semicolon if you intend to use a subcommand. The subcommand must end with a period.

Incidentally, if you omit the subcommand (we are 100% confident that you will eventually make this mistake), Minitab will generate random numbers from a standard normal distribution (i.e., mean = 0 and standard deviation = 1).

Note: Readers who will be using a student version of Minitab should read the following. All others should skip to the subsection titled Column Commands.

Student versions of Minitab do not allow the use of the subcommand *EXPONENTIAL*. Instead, we will generate random numbers from a chi-square distribution and then multiply by a constant to yield an exponential distribution. For example, if we want to generate 1000 observations from an exponential distribution whose mean is 10 using a full PC or a mainframe version of Minitab, we would type the commands

RANDOM 1000 C1;

EXPONENTIAL 10.

Using a student version of Minitab we command

RANDOM 1000 C1;

CHISQUARE 2.

MULTIPLY C1 5 C1

These commands generate 1000 random numbers from a chi-square distribution with 2 degrees of freedom, which is equivalent to an exponential distribution whose mean is 2. (Incidentally, a chi-square distribution with 10 degrees of freedom is *not* equivalent to an exponential distribution whose mean is 10.) The command *MULTIPLY C1 5 C1* multiplies the numbers in column 1 by 5 and stores the results in column 1, replacing the original numbers. Thus, the data now in column 1 could have been produced from an exponential distribution whose mean is 10. If you did not fully comprehend our perfectly lucid explanation, don't worry. The enclosed diskette contains all the required commands, so you don't have to type the commands yourself.

Column Commands

The command *HISTOGRAM C1* instructs Minitab to draw the histogram of the data in column 1. We can specify the midpoint of the first interval K1 and the width of intervals K2 as follows:

HISTOGRAM C1 K1 K2

To calculate various statistics, use

DESCRIBE C1

If you only want to see the mean of the data in column 1, type

MEAN C1

If you only want to compute the standard deviation, type

STDEV C1

We can use the *LET* command to yield various computations. For example,

LET C2=C1**2

squares the numbers in column 1 and stores the squares in column 2. Other uses of the *LET* command are as follows:

LET C3 = C1 + C2

LET C3 = C1 – C2

LET C3 = C1 * C2

LET C3 = C1 / C2

Coding Data

To help count the number of times certain events occur, we will occasionally have to code the data. For example, the command

CODE (–99:–1.96) –1 (–1.96:1.96) 0 (1.96:99) 1 C20 C21

tells Minitab to peruse the data in column 20. Any observation that lies between –99 and –1.96 is to be assigned a value of –1, observations between –1.96 and 1.96 are coded 0, and numbers between 1.96 and 99 are coded 1. The coded values are stored in column 21. To produce the number of values in each category, we will then type

TABLE C21

Row Commands

As you will soon see the experiments described in this manual require that you repeat statistical procedures many times. For example, to understand how a sampling distribution is created, you will generate random samples from specific distributions and, for each sample, compute statistics such as the mean and the standard deviation.

There are several ways to accomplish this task. The simplest would be to generate one sample at a time and calculate the necessary statistic. For instance, the following commands

RANDOM 20 C1;

NORMAL 50 10.

MEAN C1

would generate a sample of 20 observations from a normal population whose mean and standard deviation are 50 and 10, respectively. The sample mean of the 20 observations is then printed. However, to generate a large number of samples, you would have to retype the same three commands again and again. Fortunately, there are easier ways to produce the desired result. For most experiments we will be able to generate a large number of sample statistics by using Minitab's row commands. To illustrate, consider the following command and subcommand.

RANDOM 20 C1–C5;

NORMAL 50 10.

These commands produce 5 columns, each containing 20 observations from a normal population with mean 50 and standard deviation 10. If you type

PRINT C1–C5

you will see an array of 20 rows and 5 columns of numbers. We can interpret this to represent 5 samples of 20 observations. This would be the traditional interpretation because Minitab, like most statistical computer packages, performs statistical techniques on columns of data. However, we can also interpret the array to represent 20 samples of 5 observations, where each row stores the data from a different sample. Because Minitab is capable of performing row operations, this interpretation is far more convenient for our purposes. For example, the command

RMEAN C1–C5 C6

or more simply

RMEAN C1–C6

computes the mean of the 5 observations in each row and stores it in column 6. If there are 20 rows of data, column 6 will contain 20 sample means. We can then use column commands such as HISTOGRAM C6, DESCRIBE C6, or MEAN C6 to describe the distribution of sample means.

There are several row commands that we will use in this manual. They are

RMEAN

RSTDEV

RMEDIAN

RSUM

RSSQ

which calculate the mean, standard deviation, median, sum, and sum of squares, respectively, of the row of data.

All the experiments in Chapters 3–6 and 8 will use this method of computing and describing sample statistics. The experiments in Chapters 2 and 7 require only column commands.

Stored Commands

Minitab allows you to store a set of commands. This will enable you to repeat the commands easily, as many times as you want. We begin by typing

>STORE 'filename'

The required commands then follow. When all the commands have been input, type

>END

Minitab will store all the commands between *STORE 'filename'* and *END*. To execute the commands, type

>EXECUTE 'filename'

and the results from the experiment will be outputted. If you type

>EXECUTE 'filename' 100

the program will automatically run 100 times.

The stored commands can be saved and used at another session. In fact this is precisely what we have done. On the diskette included with this manual, we have stored the commands that are used in each experiment. To recall and execute the stored commands for Experiment 2.1, for example, proceed as follows: Place the diskette in drive A (we assume you will be using drive A) and type

>EXECUTE 'A:EXP2-1'

The computer will show the stored commands and then execute them.

The equivalent stored programs for the student version of Minitab will be run using the command

>EXECUTE 'A:EXP2-1S'

Users of mainframe computer versions of Minitab can also use the stored commands. Have your computer center upload the contents of the diskette into a file. For example, at our university the diskette was uploaded into a file titled /usr2/bigfiles/. To execute the commands described in Experiments 2.1, we type

>EXECUTE '/usr2/bigfiles/EXP2-1.MTB'

Your computer center can provide specific details on how to execute the experiments on your mainframe computer.

FORMAT AND PROCEDURES

In general we will discuss a concept briefly and then describe the experiment designed to illustrate that concept. For each experiment, we present its objective and discuss how the experiment is conducted. We also list the Minitab commands and provide instructions on how to execute the stored program from the diskette (including the student version of Minitab). After performing the experiment, you are expected to write a brief report of your discoveries. A form listing the relevant questions is provided for each experiment at the end of each chapter. The report can be removed and submitted to your instructor.

CHAPTER 2

Descriptive Statistics

2.1 GRAPHICAL METHODS

The purpose of descriptive statistics is to summarize data so that the meaningful essentials can be extracted. Graphical descriptive techniques fulfill this function by drawing a picture or graph of the data. The two most commonly used graphical methods are the histogram and the boxplot.

In the usual application of statistics, we start with a set of data and apply statistical techniques in order to learn something about the population from which the data were taken. In this chapter we reverse that process. In order to learn more about how descriptive statistical techniques work, we will generate the data ourselves from a known distribution. Then we'll employ statistical techniques to illustrate how the techniques work.

In the experiments described in this manual, we will generate random numbers from several different distributions. The three distributions most commonly used in this manual are the normal, the uniform, and the exponential. Figures 2.1—2.3 describe these distributions. We chose to use these distributions in this manual because, in many of the inferential statistical techniques that you are taught, one of the required conditions for the validity of the results is that the population be normal. We intend to show—or more precisely, you will discover—what happens when the population is normal, when it is somewhat nonnormal, and when it is extremely nonnormal. As you can see from Figures 2.1—2.3, the exponential distribution is extremely nonnormal, but the uniform distribution is only moderately nonnormal. The experiments described in this chapter will give you an opportunity to study these distributions.

EXPERIMENT 2.1

Objective: To show how graphical techniques describe data drawn from different populations.

Experiment: We will generate 1000 observations from a normal, a uniform, and an exponential distribution. For each set of data, Minitab will draw a histogram and a boxplot.

Minitab commands:

> *ERASE C1–C49*
>
> *RANDOM 1000 C1;*

NORMAL 25 10.

RANDOM 1000 C2;

UNIFORM 0 50.

RANDOM 1000 C3;

EXPONENTIAL 25.

HISTOGRAM C1

BOXPLOT C1

HISTOGRAM C2

BOXPLOT C2

HISTOGRAM C3

BOXPLOT C3

Notice that we start by erasing all the columns of the worksheet that we may use. (Because some older versions of Minitab are limited to 49 columns, our experiments will not go beyond this limitation.) This will ensure that we can use all the storage available in our worksheet.

Instructions for users of full PC versions of Minitab: Type

EXECUTE 'A:EXP2–1'

Instructions for users of student versions of Minitab: Type

EXECUTE 'A:EXP2–1S'

As we discussed in Chapter 1, student versions use the *CHISQUARE* subcommand to generate exponentially distributed data. Also note that the number of observations has been decreased to 500.

Use the form provided at the end of this chapter to write your report.

FIGURE 2.1   NORMAL DISTRIBUTION

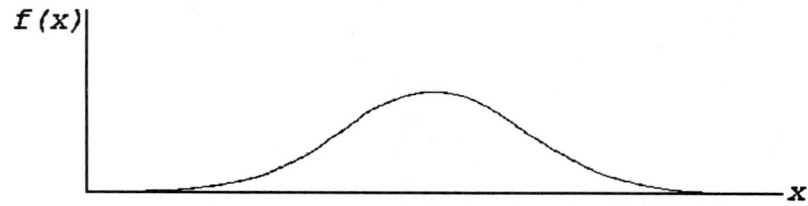

FIGURE 2.2   UNIFORM DISTRIBUTION

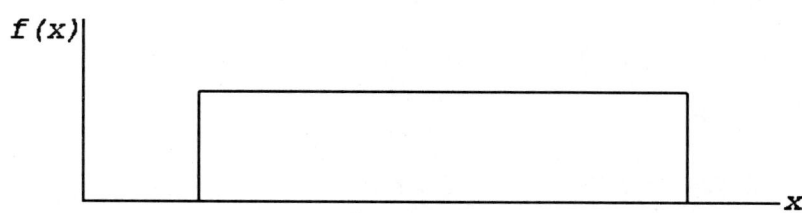

FIGURE 2.3   EXPONENTIAL DISTRIBUTION

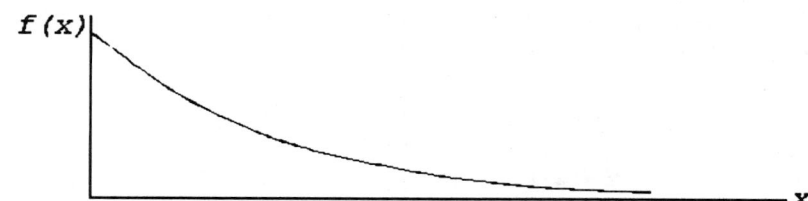

2.2 NUMERICAL DESCRIPTIVE STATISTICS

There are several different types of numerical descriptive statistics. Measures of central location describe the approximate center of the data to gauge where the observations are located. The most commonly used measures of central location are the mean and the median. Measures of dispersion judge the degree to which the data are spread out. The variance and the standard deviation are the most popular measures of dispersion. Measures of relative position allow statisticians to determine the position of a particular number relative to the others in the data set. Percentiles and quartiles are the most frequently used such measures. This section describes an experiment that should provide some insight into how these numerical measures complete their tasks.

EXPERIMENT 2.2

Objective: To demonstrate how numerical descriptive techniques describe data from three different populations.

Experiment: Minitab generates 1000 observations from each of a normal, a uniform, and an exponential distribution and calculates various statistics.

Minitab commands:

> *ERASE C1–C49*
>
> *RANDOM 1000 C1;*
>
> *NORMAL 25 10.*
>
> *RANDOM 1000 C2;*
>
> *UNIFORM 0 50.*
>
> *RANDOM 1000 C3;*
>
> *EXPONENTIAL 25.*
>
> *DESCRIBE C1 C2 C3*

Instructions for users of full PC versions of Minitab: Type

> *EXECUTE 'A:EXP2–2'*

Instructions for users of student versions of Minitab: Type

> *EXECUTE 'A:EXP2–2S'*

REPORT FOR EXPERIMENT 2.1

Do the histograms look the way you expected them to look?

Why don't the histograms appear exactly as the distributions graphed in Figures 2.1, 2.2, and 2.3?

How do the boxplots work to indicate that the three sets of data are different from one another?

REPORT FOR EXPERIMENT 2.2

Are the means and standard deviations computed in Experiment 2.2 what you had anticipated? Explain.

What do the descriptive statistics tell you about the differences among the three data sets?

CHAPTER 3

Sampling Distributions

3.1 INTRODUCTION

Conceptually, the most important component of statistical inference is the sampling distribution. To grasp the idea of a sampling distribution, consider the population created by throwing a fair die infinitely many times with the random variable x indicating the number of spots showing on any one throw. The probability distribution of the random variable x is

x	$p(x)$
1	1/6
2	1/6
3	1/6
4	1/6
5	1/6
6	1/6

The population is infinitely large, since we can throw the die infinitely many times (or at least imagine doing so). From the definitions of expectation and variance, we can show that the mean is $\mu = 3.5$, the variance is $\sigma^2 = 2.92$, and the standard deviation is $\sigma = 1.71$. Now pretend that μ is unknown and that we wish to estimate its value by using the sample mean $\bar{x}$, calculated from a sample of size $n = 2$. In actual practice, only one sample would be drawn, and hence there would be only one value of $\bar{x}$; but in order to assess how closely $\bar{x}$ estimates the value of μ, we will develop the sampling distribution of $\bar{x}$.

Consider all of the possible different samples of size 2 that could be drawn from the parent population. For each possible sample, we can calculate the mean as shown in Table 3.1. Since the value of the sample mean $\bar{x}$ varies randomly from sample to sample, we can regard $\bar{x}$ as a new random variable created by sampling. Table 3.1 lists all the possible samples and their corresponding values of $\bar{x}$.

There are 36 different possible samples of size 2, and since each sample is equally likely, the probability of any one sample's being selected is 1/36. However, $\bar{x}$ can assume only 11 different possible values: 1.0, 1.5, 2.0, . . . ,6.0, with certain values of $\bar{x}$ occurring more frequently than others. The value $\bar{x} = 1.0$ occurs only once, so its probability is 1/36. The value $\bar{x} = 1.5$ can occur in two ways; hence, $p(1.5) = 2/36$. The probabilities of the other values of $\bar{x}$ are determined similarly, and the sampling distribution of $\bar{x}$ that results is shown in Table 3.2. This same distribution can be approximated by tossing two fair dice many times; the distribution of $\bar{x}$ thus created should be similar to the one shown in Table 3.2.

TABLE 3.1 ALL SAMPLES OF SIZE 2 AND THEIR MEANS

Sample	$\bar{x}$	Sample	$\bar{x}$	Sample	$\bar{x}$
1, 1	1.0	3, 1	2.0	5, 1	3.0
1, 2	1.5	3, 2	2.5	5, 2	3.5
1, 3	2.0	3, 3	3.0	5, 3	4.0
1, 4	2.5	3, 4	3.5	5, 4	4.5
1, 5	3.0	3, 5	4.0	5, 5	5.0
1, 6	3.5	3, 6	4.5	5, 6	5.5
2, 1	1.5	4, 1	2.5	6, 1	3.5
2, 2	2.0	4, 2	3.0	6, 2	4.0
2, 3	2.5	4, 3	3.5	6, 3	4.5
2, 4	3.0	4, 4	4.0	6, 4	5.0
2, 5	3.5	4, 5	4.5	6, 5	5.5
2, 6	4.0	4, 6	5.0	6, 6	6.0

TABLE 3.2 SAMPLING DISTRIBUTION OF $\bar{x}$

$\bar{x}$	$p(\bar{x})$
1.0	1/36
1.5	2/36
2.0	3/36
2.5	4/36
3.0	5/36
3.5	6/36
4.0	5/36
4.5	4/36
5.0	3/36
5.5	2/36
6.0	1/36

The most interesting aspect of the sampling distribution of $\bar{x}$ is how dissimilar it is from the distribution of x, as can be seen in Figures 3.1 and 3.2. Figure 3.1 depicts the distribution of x, and Figure 3.2 depicts the sampling distribution of $\bar{x}$. We can also see the differences between the distributions by calculating the mean and variance of the sampling distribution. Using the rules of expectation and variance, we find $\mu_{\bar{x}} = 3.5$ and $\sigma_{\bar{x}}^2 = 1.46$. From our previous calculations, we found that the mean of the population is $\mu = 3.5$ and that the variance is $\sigma^2 = 2.92$. It is no coincidence that the means are equal and that the variance of $\bar{x}$ is exactly half the variance of x, as we shall see shortly.

If we now repeat the sampling process with the same population but with other values of n, we produce somewhat different sampling distributions of $\bar{x}$. Figure 3.3 shows the sampling distributions of $\bar{x}$ when $n = 5$, 10, and 25. As n grows larger, the number of possible values of $\bar{x}$ also grows larger; consequently, the histograms depicted in Figure 3.3 have been smoothed (to avoid drawing a large number of rectangles). Observe that, in each case, $\mu_{\bar{x}} = \mu$ and $\sigma_{\bar{x}}^2 = \sigma^2/n$.

Another thing that happens as n gets larger is that the sampling distribution of $\bar{x}$ becomes increasingly bell-shaped. This phenomenon is summarized in the Central Limit Theorem.

CENTRAL LIMIT THEOREM

If relatively large samples of size n are drawn from any population, the sampling distribution of $\bar{x}$ is approximately normal. The approximation is better for larger sample sizes.

We can now summarize what we know about the sampling distribution of the sample mean.

Sampling Distribution of $\bar{x}$

1. $\bar{x}$ is approximately normally distributed
2. $\mu_{\bar{x}} = \mu$
3. $\sigma_{\bar{x}}^2 = \sigma^2/n$

The accuracy of the approximation alluded to in the Central Limit Theorem depends on the probability distribution of the parent population and on the sample size. If the population is normal, then $\bar{x}$ is normally distributed for all values of n. If the population is nonnormal, then $\bar{x}$ is approximately normal only for larger values of n. In Figure 3.3, the distribution of $\bar{x}$ starts looking normal when $n > 10$. In many practical situations, a sample size of $n > 30$ may be sufficiently large to allow us to use the normal distribution as an approximation for the sampling distribution of $\bar{x}$. We urge you, however, to be cautious about the sample size. If a population is quite nonnormal, the sampling distribution will also be nonnormal—even for moderately large values of n.

FIGURE 3.1   DISTRIBUTION OF *x*

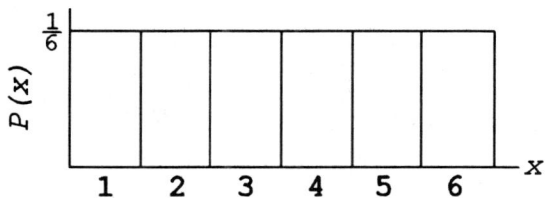

FIGURE 3.2   DISTRIBUTION OF $\bar{x}$

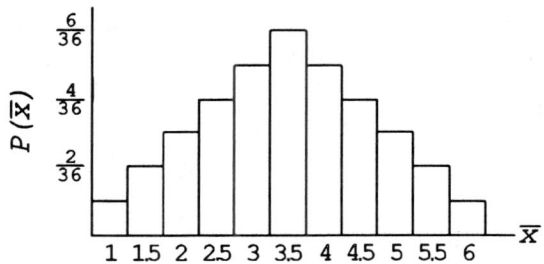

FIGURE 3.3   SAMPLING DISTRIBUTIONS OF $\bar{x}$ WHEN n = 5, 10 AND 25

n = 5
$\mu_{\bar{x}}$ = 3.5
$\sigma_{\bar{x}}^2$ = .5833 (= $\sigma^2/5$)

n = 10
$\mu_{\bar{x}}$ = 3.5
$\sigma_{\bar{x}}^2$ = .2917 (= $\sigma^2/10$)

n = 25
$\mu_{\bar{x}}$ = 3.5
$\sigma_{\bar{x}}^2$ = .1167 (= $\sigma^2/25$)

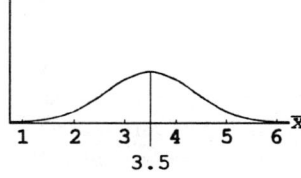
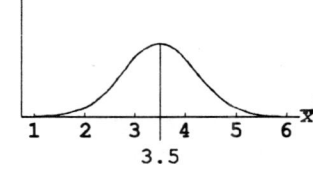
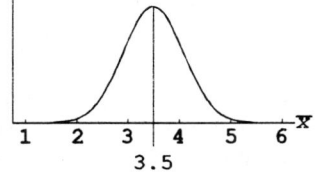

3.2 CREATING THE SAMPLING DISTRIBUTION OF THE MEAN

Minitab can be used to demonstrate how sampling distributions are created, what they look like, and what their parameters are. Let's begin with the sampling distribution of the mean of a sample of two tosses of a fair die described theoretically above. The commands

> *RANDOM 100 C1;*

> *INTEGER 1 6.*

store 100 numbers in column 1 where the numbers are drawn from a distribution of integers between 1 and 6. Thus, column 1 now contains numbers that could have been observed from 100 tosses of a fair die. To examine this claim, type

> *PRINT C1*

and

> *HISTOGRAM C1*

The second command should produce something like Figure 3.1. That is, each of the six values occurs with approximately the same frequency. The command

> *DESCRIBE C1*

will output the mean and standard deviation of our 100 observations. You should observe that these numbers are similar to the theoretical mean and standard deviation of the distribution of x ($\mu = 3.5$ and $\sigma = 1.71$).

At this point it's probably worthwhile reiterating a point we made in Chapter 1. What we are doing here is simulating throwing a fair die 100 times. The distribution and the mean and standard deviation of these 100 values will not be exactly the same as the theoretical distribution. In order for our observed values to exactly match the theoretical distribution, we would have to generate the entire population, an impossible task because the population is infinitely large. If we were to generate 1000 values instead of only 100, it is quite likely that the observed distribution would more closely resemble the theoretical distribution. You can prove this to yourself by typing

> *RANDOM 1000 C1;*

> *INTEGER 1 6.*

> *HISTOGRAM C1*

> *DESCRIBE C1*

Now that you understand the difference between an observed distribution and a theoretical distribution, we can proceed to demonstrate how the sampling distribution of the mean is generated.

EXPERIMENT 3.1

Objective: To simulate tossing two dice 1000 times and calculate the distribution of the mean of the two tosses.

Experiment: Minitab generates 1000 integers between 1 and 6 in each of columns 1 and 2. We use a row command to calculate the mean of the two integers in each of the 1000 rows. Each row represents a sample of size 2. The histogram will be drawn and descriptive statistics calculated.

Minitab commands:

> *ERASE C1–C49*
>
> *RANDOM 1000 C1–C2;*
>
> *INTEGER 1 6.*
>
> *RMEAN C1–C3*
>
> *HISTOGRAM C3*
>
> *DESCRIBE C3*

Instructions for all users of Minitab: Type

> *EXECUTE 'A:EXP3–1'*

EXPERIMENT 3.2

Objective: To develop the sampling distribution of the mean when we toss five dice.

Experiment: Minitab will generate 1000 integers between 1 and 6 in each of columns 1 to 5. Each row represents a sample of size 5. The row command *RMEAN* computes the sample mean for each row. Minitab then draws the histogram and outputs descriptive statistics.

Minitab commands:

> *ERASE C1–C49*
>
> *RANDOM 1000 C1–C5;*
>
> *INTEGER 1 6.*
>
> *RMEAN C1–C6*
>
> *HISTOGRAM C6*
>
> *DESCRIBE C6*

Instructions for users of full PC versions of Minitab: Type

> *EXECUTE 'A:EXP3–2'*

Instructions for users of student versions of Minitab: Type

> *EXECUTE 'A:EXP3–2S'*

The number of samples generated is decreased to 500.

3.3 SAMPLING DISTRIBUTION OF THE MEAN: NORMAL AND NONNORMAL POPULATIONS

We know from our discussion above that if the population is normal, the sampling distribution of the mean is normal. If the population is nonnormal, the sample mean will be approximately normally distributed provided that the sample size is sufficiently large. The size of the sample required to make $\bar{x}$ approximately normally distributed depends on how nonnormal the population is. This issue will be the focus of the next eight experiments.

EXPERIMENT 3.3

Objective: To examine the sampling distribution of the mean when the population is normal.

Experiment: Minitab will generate 1000 samples of size 2 from a normal population. As before, the rows will represent the 1000 samples (in this case, the sample size is 2). The *RMEAN* command will compute the mean for each sample, and the sampling distribution will be drawn and described.

Minitab commands:

    ERASE C1–C49

    RANDOM 1000 C1–C2;

    NORMAL 50 10.

    RMEAN C1–C3

    HISTOGRAM C3

    DESCRIBE C3

Instructions for all users of Minitab: Type

    EXECUTE 'A:EXP3–3'

EXPERIMENT 3.4

Objective: To examine the effect, on the sampling distribution of the mean, of increasing the sample size when the population is normal.

Experiment: We repeat Experiment 3.3 increasing the sample size to 25. (We reduce the number of samples generated to 500 because of worksheet limitations.)

Minitab commands:

    ERASE C1–C49

    RANDOM 500 C1–C25;

    NORMAL 50 10.

    RMEAN C1–C26

    HISTOGRAM C26

    DESCRIBE C26

Instructions for users of full PC versions of Minitab: Type

    EXECUTE 'A:EXP3–4'

Instructions for users of student versions of Minitab: Type

EXECUTE 'A:EXP3–4S'

The number of samples to be generated is reduced to 100.

EXPERIMENT 3.5

Objective: To see what happens to the sampling distribution of the mean if the population is moderately nonnormal.

Experiment: Minitab will generate samples of size 2 from a uniform distribution that ranges from 0 to 10. For each sample, we compute the sample mean, which allows us to draw the sampling distribution, and calculate descriptive statistics.

Minitab commands:

ERASE C1–C49

RANDOM 1000 C1–C2;

UNIFORM 0 10.

RMEAN C1–C3

HISTOGRAM C3

DESCRIBE C3

Instructions for all users of Minitab: Type

EXECUTE 'A:EXP3–5'

EXPERIMENT 3.6

Objective: To examine the sampling distribution of the mean when the population is nonnormal, and, in concert with Experiment 3.5, to determine the effect, on the sampling distribution of the mean, of increasing the sample size when the population is moderately nonnormal.

Experiment: We repeat Experiment 3.5 increasing the sample size to 10.

Minitab commands:

    ERASE C1–C49

    RANDOM 1000 C1–C10;

    UNIFORM 0 10.

    RMEAN C1–C11

    HISTOGRAM C11

    DESCRIBE C11

Instructions for users of full PC versions of Minitab: Type

    EXECUTE 'A:EXP3–6'

Instructions for users of student versions of Minitab: Type

    EXECUTE 'A:EXP3–6S'

The number of samples to be generated is reduced to 200.

EXPERIMENT 3.7

Objective: Together with Experiments 3.5 and 3.6, to illustrate the effect, on the sampling distribution of the mean, of increasing the sample size when the population is moderately nonnormal.

Experiment: We repeat Experiment 3.5 increasing the sample size to 25 (the number of samples is decreased to 500).

Minitab commands:

    ERASE C1–C49

    RANDOM 500 C1–C25;

    UNIFORM 0 10.

    RMEAN C1–C26

HISTOGRAM C26

DESCRIBE C26

Instructions for users of full PC versions of Minitab: Type

EXECUTE 'A:EXP3-7'

Instructions for users of student versions of Minitab: Type

EXECUTE 'A:EXP3-7S'

The number of samples to be generated is reduced to 100.

EXPERIMENT 3.8

Objective: To develop the sampling distribution of the mean when the sample size is small and the population is extremely nonnormal.

Experiment: Minitab will generate 1000 samples of size 2 from an exponential distribution and compute the mean for each sample. The sampling distribution of the mean will then be drawn and described.

Minitab commands:

ERASE C1-C49

RANDOM 1000 C1-C2;

EXPONENTIAL 10.

RMEAN C1-C3

HISTOGRAM C3

DESCRIBE C3

Instructions for users of full PC versions of Minitab: Type

EXECUTE 'A:EXP3-8'

Instructions for users of student versions of Minitab: Type

EXECUTE 'A:EXP3-8S'

EXPERIMENT 3.9

Objective: To examine the effect, on the sampling distribution of the mean, of increasing the sample size when the population is extremely nonnormal.

Experiment: We repeat Experiment 3.8 using a sample size of 10.

Minitab commands:

    ERASE C1–C49

    RANDOM 1000 C1–C10;

    EXPONENTIAL 10.

    RMEAN C1–C11

    HISTOGRAM C11

    DESCRIBE C11

Instructions for users of full PC versions of Minitab: Type

    EXECUTE 'A:EXP3–9'

Instructions for users of student versions of Minitab: Type

    EXECUTE 'A:EXP3–9S'

The number of samples to be generated is reduced to 200.

EXPERIMENT 3.10

Objective: In concert with Experiments 3.8 and 3.9, to determine the effect of increasing the sample size, on the sampling distribution of the mean, when the population is extremely nonnormal.

Experiment: Repeat Experiment 3.8 using a sample size of 25.

Minitab commands:

    ERASE C1–C49

    RANDOM 500 C1–C25;

EXPONENTIAL 10.

RMEAN C1–C26

HISTOGRAM C26

DESCRIBE C26

Instructions for users of full PC versions of Minitab: Type

EXECUTE 'A:EXP3–10'

Instructions for users of student versions of Minitab: Type

EXECUTE 'A:EXP3–10S'

The number of samples to be generated is reduced to 100.

3.4   SAMPLING DISTRIBUTIONS OF OTHER STATISTICS

The sampling distribution of the mean is only one of many sampling distributions encountered in a typical statistics course. We can develop the sampling distribution for any sample statistic. For example, to generate the sampling distribution of the median, we use the command *RMEDIAN*, which computes the medians of the rows (rather than the means computed above).

EXPERIMENT 3.11

Objective: To develop the sampling distribution of the median when the population is normally distributed.

Experiment: We will generate 500 random samples of size 25 from a normal population whose mean is 50 and whose standard deviation is 10. For each sample, we compute the median and then draw the histogram and calculate statistics for the sampling distribution of the median.

Minitab commands:

ERASE C1–C49

RANDOM 500 C1–C25;

NORMAL 50 10.

RMEDIAN C1–C26

HISTOGRAM C26

DESCRIBE C26

Instructions for users of full PC versions of Minitab: Type

EXECUTE 'A:EXP3–11'

Instructions for users of student versions of Minitab: Type

EXECUTE 'A:EXP3–11S'

The number of samples to be generated is reduced to 100.

    Later in your statistics course you may encounter the sampling distribution of the variance. Minitab does not compute the sample variances of the rows. However, it does calculate the sample standard deviation using the command *RSTDEV* from which the variances can be computed.

EXPERIMENT 3.12

Objective: To develop the sampling distribution of the variance when the population is normal.

Experiment: Minitab will generate 500 samples of size 25 from a normal population whose mean is 50 and whose standard deviation is 10. The sample variances will be calculated and the sampling distribution of the variance developed.

Minitab commands:

ERASE C1–C49

RANDOM 500 C1–C25;

NORMAL 50 10.

RSTDEV C1–C26

*LET C27=C26**2*

HISTOGRAM C27

DESCRIBE C27

Instructions for users of full PC versions of Minitab: Type

    EXECUTE 'A:EXP3–12'

Instructions for users of student versions of Minitab: Type

    EXECUTE 'A:EXP3–12S'

The number of samples to be generated is reduced to 100.

REPORT FOR EXPERIMENT 3.1

Compare the histogram that was drawn in this experiment with the theoretical sampling distribution shown in Figure 3.2. Are they similar?

What did you anticipate seeing when Minitab computed the mean and standard deviation of the sampling distribution of the mean?

What did you actually observe when you computed the mean and standard deviation of the sampling distribution of the mean?

Discuss the significance of the results of this experiment.

REPORT FOR EXPERIMENT 3.2

What values for the mean and standard deviation did you anticipate seeing?

Value of the observed mean of the sampling distribution:

Value of the observed standard deviation of the sampling distribution of the mean:

Compare the histograms drawn in Experiment 3.1 with the one drawn in this experiment. Discuss similarities and differences.

(continued)

What do the results of Experiments 3.1 and 3.2 tell you about the effect of increasing the sample size on the sampling distribution of the mean?

REPORT FOR EXPERIMENT 3.3

What are the mean and standard deviation of the population from which we're sampling?

Draw the distribution of the population from which we're sampling.

Use the Central Limit Theorem to determine the mean and standard deviation of the sampling distribution of the sample mean calculated by drawing samples of size 2 from a normal population whose mean is 50 and whose standard deviation is 10.

(*continued*)

What are the mean and standard deviation of the sampling distribution created by Experiment 3.3?

Does the histogram produced in this experiment appear to have a bell shape?

Does this experiment confirm the Central Limit Theorem and what you anticipated seeing when the mean and standard deviation of the sampling distribution of the mean were computed? Explain.

REPORT FOR EXPERIMENT 3.4

Theoretical value of the mean and standard deviation of the sampling distribution of the mean:

Observed value of the mean and standard deviation of the sampling distribution of the mean:

Does the histogram produced in this experiment appear to have a bell shape?

Do the results of this experiment confirm what the theory tells you about the sampling distribution of the mean when the population is normal? Explain.

(continued)

Compare the shapes of the histograms and the descriptive measures of the sampling distributions created in Experiments 3.3 and 3.4. Discuss the similarities and differences and explain the cause of the differences.

REPORT FOR EXPERIMENT 3.5

What are the mean and standard deviation of a uniform distribution that ranges from 0 to 10? (Calculate these parameters using the definition of expected value and variance.)

Use the Central Limit Theorem to determine the mean and standard deviation of the sampling distribution of the mean when the population is uniform ranging from 0 to 10 and the sample size is 2.

Observed value of the mean and standard deviation of the sampling distribution of the mean:

Draw the distribution of the population from which we've sampled.

(continued)

Discuss whether the histogram drawn in this experiment is bell-shaped.

Do the results of this experiment support the Central Limit Theorem? Explain.

REPORT FOR EXPERIMENT 3.6

Theoretical value of the mean and standard deviation of the sampling distribution of the sample mean:

Observed value of the mean and standard deviation of the sampling distribution of the sample mean:

Discuss whether the histogram is bell-shaped.

Describe the similarities and differences between the shapes of the histograms and the descriptive measures of the sampling distributions created in Experiments 3.5 and 3.6.

(continued)

What do the results of Experiments 3.5 and 3.6 tell you about the Central Limit Theorem?

REPORT FOR EXPERIMENT 3.7

Theoretical mean and standard deviation of the sampling distribution of the sample mean:

Observed mean and standard deviation of the sampling distribution of the sample mean:

Is this histogram more bell-shaped than those shown in Experiments 3.5 and 3.6? If so, discuss why this may be so.

(*continued*)

What do the results of Experiments 3.5—3.7 tell you about the effect of increasing the sample size, on the sampling distribution of the mean, when the population is moderately nonnormal?

REPORT FOR EXPERIMENT 3.8

Theoretical mean and standard deviation of the sampling distribution of the mean when the sample size is 2:

Observed mean and standard deviation of the sampling distribution of the sample mean:

Does the histogram appear to be bell-shaped? If not, what does it look like?

Does it appear that the sampling distribution of the mean is normal? Explain.

(continued)

What do the results of this experiment tell you about the sampling distribution of the mean when the sample size is small and the population is extremely nonnormal?

REPORT FOR EXPERIMENT 3.9

Theoretical mean and standard deviation of the sampling distribution of the sample mean when the sample size is 10:

Observed mean and standard deviation of the sampling distribution of the sample mean:

Does the histogram appear to be bell-shaped? Is it more bell-shaped than the histogram created by Experiment 3.8?

What do the results of this experiment tell you about the effect of increasing the sample size, on the sampling distribution of the mean, when the population is extremely nonnormal?

REPORT FOR EXPERIMENT 3.10

Theoretical mean and standard deviation of the sampling distribution of the sample mean when the sample size is 25:

Observed mean and standard deviation of the sampling distribution of the sample mean:

Does the histogram appear to be bell-shaped? Is it more bell-shaped than the histograms created in Experiments 3.8 and 3.9? Explain why this may be so.

What do Experiments 3.8 to 3.10 tell you about the Central Limit Theorem?

REPORT FOR EXPERIMENT 3.11

Does the histogram produced in this experiment appear to be bell-shaped?

Compare the mean and standard deviation of the sampling distribution of the median with the mean and standard deviation of the population from which we've sampled:

Compare the mean and standard deviation of the sampling distribution of the median with the mean and standard deviation of the sampling distribution of the mean (Experiment 3.4):

What conclusions can you draw regarding the sampling distribution of the median when the population is normal?

REPORT FOR EXPERIMENT 3.12

Does it appear that the sampling distribution of the variance is normal?

Does it appear that the Central Limit Theorem applies to the sampling distribution of the variance as it did to the sampling distribution of the mean? Explain.

CHAPTER 4

Estimation

4.1 INTRODUCTION

As its name suggests, the objective of estimation is to determine the approximate value of a population parameter on the basis of a sample statistic. We can use sample data to estimate a population parameter in two ways. First, we can compute the value of the estimator and consider that value as the estimate. Such an estimator is called a point estimator. In drawing inferences about a population, it is intuitively reasonable to expect that a large sample will produce more accurate results since it contains more information than a smaller sample does. But point estimators don't have the capacity to reflect the effects of larger sample sizes. The second way of estimating a population parameter is to use an interval estimator. As you will see, the interval estimator is affected by the sample size; and because of this feature the interval estimator is used far more frequently.

Numerous applications of estimation occur in the real world. For example, television network executives want to know the proportion of television viewers who are tuned into their network; a production manager wishes to know the average daily production in his plant; a union negotiator would like to know the average annual income of North American blue-collar workers. In each of these cases, in order to accomplish the objective exactly, the interested party would have to examine each member of their population and then calculate the parameter of interest. For instance, the union negotiator would have to ask every North American blue-collar worker what his or her annual income is, and then calculate the average of these values—a task that is both impractical and prohibitively expensive. An alternative would be to take a random sample from this population, calculate the sample mean, and use that as an estimator of the population mean. The use of the sample mean to estimate the population mean seems logical. The selection of the sample statistic to be used as an estimator, however, depends on the characteristics of that statistic. Naturally, we want to use the statistic with the most desirable qualities for our purposes. One such desirable quality of an estimator is unbiasedness.

Unbiasedness

An unbiased estimator of a population parameter is one whose expected value is equal to that parameter.

This means that, if you were to take an infinite number of samples, calculate the value of the estimator in each sample, and then average these values, the average value would equal the parameter. Essentially this amounts to saying that, on average, the sample statistic is equal to the parameter.

Knowing that an estimator is unbiased only assures us that its expected value equals the parameter; it does not tell us how close the estimator is to the parameter. Another desirable quality is for the estimator to be as close to its parameter as possible; and certainly, as the sample size grows larger, the sample statistic should come closer to the population parameter. This quality is called consistency.

Consistency

An unbiased estimator is said to be consistent if the difference between the estimator and the parameter grows smaller as the sample size grows larger.

The measure we use to gauge closeness is the variance (or the standard deviation). Thus, $\bar{x}$ is a consistent estimator of μ, since the variance of $\bar{x}$ is σ^2/n. This implies that, as n grows larger, the variance σ^2/n grows smaller. As a consequence, an increasing proportion of the statistics $\bar{x}$ fall close to μ.

In Sections 4.2 and 4.3 we describe experiments that deal with the concepts of unbiasedness and consistency. In Sections 4.4 and 4.5 we discuss confidence interval estimators of population means, and in Section 4.6 we discuss sample size selection.

4.2   UNBIASEDNESS

As we explained earlier, an unbiased estimator of a population parameter is an estimator whose average value in repeated sampling would be equal to that parameter. The next four experiments demonstrate the concept of unbiasedness.

EXPERIMENT 4.1

Objective: To illustrate that $\bar{x}$ is an unbiased estimator of μ.

Experiment: We generate 1000 samples of size 10 from a normal population whose mean is 50. We compute the sample means and the mean of the sample means. The mean of the sample means is outputted.

Minitab commands:

    ERASE C1–C49

    RANDOM 1000 C1–C10;

    NORMAL 50 10.

RMEAN C1–C11

MEAN C11

Instructions for users of full PC versions of Minitab: Type

EXECUTE 'A:EXP4–1'

Instructions for users of student versions of Minitab: Type

EXECUTE 'A:EXP4–1S'

The number of samples to be generated is reduced to 200.

When the population is normal, the sample median is also an unbiased estimator of the population mean.

EXPERIMENT 4.2

Objective: To illustrate that the sample median is an unbiased estimator of the population mean when the population is normal.

Experiment: Minitab will generate 1000 samples of size 10 from a normal population whose mean is 50. The sample medians and the mean of the sample medians will then be computed.

Minitab commands:

ERASE C1–C49

RANDOM 1000 C1–C10;

NORMAL 50 10.

RMEDIAN C1–C11

MEAN C11

Instructions for users of full PC versions of Minitab: Type

EXECUTE 'A:EXP4–2'

Instructions for users of student versions of Minitab: Type

> *EXECUTE 'A:EXP4–2S'*

The number of samples to be generated is reduced to 200.

The next two experiments examine whether the sample variance s^2 is an unbiased estimator of the population variance σ^2. Recall that s^2 is defined in the following way:

$$\frac{\Sigma(x_i - \bar{x})^2}{n - 1}$$

When you first encountered this formula, you may have wondered why the denominator was $n - 1$ rather than the seemingly more logical value of n. We divide by $n - 1$ because by doing so we produce an estimator that is an unbiased estimator of the population variance σ^2. Experiments 4.3 and 4.4 will demonstrate this point.

EXPERIMENT 4.3

Objective: To illustrate that the sample variance s^2 is an unbiased estimator of the population variance σ^2.

Experiment: Generate 500 samples of size 4 from a normal population whose variance is 100 (standard deviation is 10). Calculate the sample variances and the mean of the sample variances.

Minitab commands:

> *ERASE C1–C49*
>
> *RANDOM 500 C1–C4;*
>
> *NORMAL 50 10.*
>
> *RSTDEV C1–C5*
>
> *LET C6=C5**2*
>
> *MEAN C6*

Note that because Minitab does not directly calculate row variances, we compute the standard deviation for each row and square that value to produce the sample variances.

Instructions for all users of Minitab: Type

> *EXECUTE 'A:EXP4–3'*

EXPERIMENT 4.4

Objective: To illustrate that the statistic

$$\frac{\Sigma(x_i - \bar{x})^2}{n}$$

is a biased estimator of the population variance.

Experiment: Minitab will generate 500 samples of size 4 from a normal population whose variance is 100 (standard deviation is 10). In each sample Minitab computes the statistic

$$\frac{\Sigma(x_i - \bar{x})^2}{n}$$

and prints the mean value of the 500 samples.

Minitab commands:

> *ERASE C1–C49*
>
> *RANDOM 500 C1–C4;*
>
> *NORMAL 50 10.*
>
> *RSUM C1–C5*
>
> *RSSQ C1–C4 C6*
>
> *LET C7=(C6–C5*C5/4)/4*
>
> *MEAN C7*

The command *RSUM C1-C5* calculates the sums of the numbers in columns 1 to 4 (which were generated from a normal population whose mean is 50 and whose standard deviation is 10) and stores them in column 5. *RSSQ C1-C4 C6* computes the sums of squares of the values in columns 1 to 4 and stores them in column 6. You should confirm that the commands

RSUM C1-C5

RSSQ C1-C4 C6

LET C7=(C6-C5*C5/4)/4

calculate

$$\frac{\Sigma(x_i - \bar{x})^2}{n} = \frac{\Sigma x_i^2 - (\Sigma x_i)^2/n}{n}$$

for each sample. The command

MEAN C7

outputs the mean of the 500 values of the newly created statistic.

Instructions for all users of Minitab: Type

EXECUTE 'A:EXP4-4'

4.3   CONSISTENCY

A consistent estimator is one whose difference between it and the parameter grows smaller as the sample size grows larger.

EXPERIMENT 4.5

Objective: Together with Experiment 4.6, to illustrate that the sample mean $\bar{x}$ is a consistent estimator of the population mean μ.

Experiment: Generate 500 samples of size 4 from a normal population whose mean and standard deviation are 20 and 5, respectively. The sampling distribution of the mean is created, graphed, and described.

Minitab commands:

ERASE C1-C49

RANDOM 500 C1-C4;

NORMAL 20 5.

RMEAN C1–C5

HISTOGRAM C5

DESCRIBE C5

Instructions for all users of Minitab: Type

EXECUTE 'A:EXP4–5'

EXPERIMENT 4.6

Objective: In concert with Experiment 4.5, to illustrate that the sample mean is a consistent estimator of the population mean.

Experiment: Minitab will generate 500 samples of size 25 from a normal population whose mean is 20 and whose standard deviation is 5. The sampling distribution of the mean is created, graphed, and described.

Minitab commands:

ERASE C1–C49

RANDOM 500 C1–C25;

NORMAL 20 5.

RMEAN C1–C26

HISTOGRAM C26

DESCRIBE C26

Instructions for users of full PC versions of Minitab: Type

EXECUTE 'A:EXP4–6'

Instructions for users of student versions of Minitab: Type

EXECUTE 'A:EXP4–6S'

The number of samples to be generated is reduced to 100.

4.4 ESTIMATING THE POPULATION MEAN WHEN THE POPULATION VARIANCE IS KNOWN

From the sampling distribution of the mean, we can derive the confidence interval estimator of μ. It is

$$\bar{x} \pm z_{\alpha/2} \frac{\sigma}{\sqrt{n}}$$

One of the fundamental concepts of statistical inference is that the confidence interval estimate does not always include the value of the parameter we're trying to estimate. That means, for example, that the process that produces 95% confidence intervals will produce intervals that contain the true value of the parameter 95% of the time. The remaining 5% of the time the interval estimate will be incorrect.

EXPERIMENT 4.7

Objective: To show that confidence interval estimates are correct most, but not all, of the time.

Experiment: We will generate 1000 samples of 9 observations from a normal population with mean 10 and standard deviation 3. For each sample we will compute the 90% confidence interval estimator, which is

$$\bar{x} \pm 1.645 \frac{\sigma}{\sqrt{n}}$$

With $\sigma = 3$ and $n = 9$ it follows that the interval estimator is

$$\bar{x} \pm 1.645 \frac{3}{\sqrt{9}} = \bar{x} \pm 1.645$$

Minitab will then be instructed to count the number of intervals that include the true value of the population mean, which is $\mu = 10$.

Minitab commands:

    ERASE C1-C49

    RANDOM 1000 C1-C9;

    NORMAL 10 3.

    RMEAN C1-C10

$$LET\ C11 = C10 - 1.645$$

$$LET\ C12 = C10 + 1.645$$

$$CODE\ (-99:10)1\ (10:99)2\ C11{-}C12\ C13{-}C14$$

$$TABLE\ C13\ C14$$

Note that columns 11 and 12 contain the lower and upper confidence limits, respectively. The code command codes each value in columns 11 and 12 and puts the results in columns 13 and 14. If a number in columns 11 and 12 is less than or equal to 10 (actually between −99 and 10; we had to specify some lower limit), the coded value is 1. If a confidence limit is 10 or more (actually between 10 and 99), the coded value is 2. If you want to confirm that this has been done, print columns 11 to 14. If a lower confidence limit is 10 or less (code = 1) and the upper confidence limit is 10 or more (code = 2), the confidence interval estimate contains the value of the population mean, 10. Look for samples where column 13 contains a 1 and column 14 contains a 2; these samples produced confidence interval estimates that included the true mean value. The command *TABLE C13 C14* is an easier way to count the number of intervals containing 10. Simply determine the number of times the number in column 13 is a 1 and the number of times the number in column 14 is a 2.

Instructions for users of full PC versions of Minitab: Type

$$EXECUTE\ 'A{:}EXP4{-}7'$$

Instructions for users of student versions of Minitab: Type

$$EXECUTE\ 'A{:}EXP4{-}7S'$$

The number of samples to be generated is reduced to 200. We recommend that you run *EXP4–7S* five times and combine the results.

EXPERIMENT 4.8

Objective: To demonstrate what happens when the confidence level is raised to 95%.

Experiment: We repeat Experiment 4.7 using a confidence level of 95%. The 95% confidence interval estimator is

$$\bar{x} \pm z_{\alpha/2}\frac{\sigma}{\sqrt{n}} = \bar{x} \pm 1.96\frac{3}{\sqrt{9}} = \bar{x} \pm 1.96$$

Minitab commands:

> *ERASE C1–C49*
>
> *RANDOM 1000 C1–C9;*
>
> *NORMAL 10 3.*
>
> *RMEAN C1–C10*
>
> *LET C11=C10 – 1.96*
>
> *LET C12=C10 + 1.96*
>
> *CODE (–99:10)1(10:99)2 C11–C12 C13–C14*
>
> *TABLE C13 C14*

Instructions for users of full PC versions of Minitab: Type

> *EXECUTE 'A:EXP4–8'*

Instructions for users of student versions of Minitab: Type

> *EXECUTE 'A:EXP4–8S'*

The number of samples to be generated is reduced to 200. We recommend that you repeat this experiment five times and combine results.

4.5 USING THE STUDENT t DISTRIBUTION TO ESTIMATE A POPULATION MEAN

Statistics professors almost always use estimating the population mean when the population variance is known to demonstrate the first example of estimation. Although assuming σ^2 is known is quite unrealistic, we use this example because it allows us to show how the sampling distribution of the mean is applied to producing the confidence interval estimator of μ. It is much more realistic to believe that if the population mean is unknown, so is the population variance. As a consequence, we must develop another sampling distribution—one that uses the sample variance rather than the population variance. The quantity

$$t = \frac{\bar{x} - \mu}{s/\sqrt{n}}$$

is said to be Student t distributed with $n - 1$ degrees of freedom, provided that the population from which we're sampling is normally distributed. The purpose of the next two experiments is to demonstrate what the Student t distribution looks like and that it is similar but more spread out than the standard normal distribution.

EXPERIMENT 4.9

Objective: To compare the standard normal and Student t sampling distributions.

Experiment: We will generate 1000 samples of size 4 from a normal population whose mean and standard deviation are 50 and 10, respectively. For each sample, we will compute both the z-statistic

$$z = \frac{\bar{x} - \mu}{\sigma/\sqrt{n}} = \frac{\bar{x} - 50}{10/\sqrt{4}} = \frac{\bar{x} - 50}{10/2}$$

and the t-statistic

$$t = \frac{\bar{x} - \mu}{s/\sqrt{n}} = \frac{\bar{x} - 50}{s/\sqrt{4}} = \frac{\bar{x} - 50}{s/2}$$

We then graph and describe both sampling distributions.

Minitab commands:

```
ERASE C1–C49

RANDOM 1000 C1–C4;

NORMAL 50 10.

RMEAN C1–C5

RSTDEV C1–C4 C6

LET C7=(C5–50)/(10/2)

LET C8=(C5–50)/(C6/2)

HISTOGRAM C7 –5 1

HISTOGRAM C8 –5 1
```

DESCRIBE C7

DESCRIBE C8

Note that column 5 will contain the sample means and column 6 will store the sample standard deviations. The values of

$$\frac{\bar{x} - \mu}{\sigma/\sqrt{n}}$$

will be in column 7 and the values of

$$\frac{\bar{x} - \mu}{s/\sqrt{n}}$$

will be in column 8.

Also notice that we have specified the midpoint of the first interval (–5) and the width of the intervals (1), making it easier to compare histograms.

Instructions for users of full PC versions of Minitab: Type

EXECUTE 'A:EXP4-9'

Instructions for users of student versions of Minitab: Type

EXECUTE 'A:EXP4-9S'

The number of samples to be generated is reduced to 400.

In preparation for the next experiment, use a Student t Table to find the probability that a Student t distributed random variable with 3 degrees of freedom is greater than 1.638. Find the probability that a standard normal random variable is greater than 1.638. What do these two probabilities tell you about the spreads of the two distributions?

EXPERIMENT 4.10

Objective: To compare the spreads of the standard normal and Student t distributions.

Experiment: We will repeat Experiment 4.9 and count the number of values of t that are greater than 1.638 and the number of values of z that are greater than 1.638.

Minitab commands:

> *ERASE C1–C49*
>
> *RANDOM 1000 C1–C4;*
>
> *NORMAL 50 10.*
>
> *RMEAN C1–C5*
>
> *RSTDEV C1–C4 C6*
>
> *LET C7=(C5–50)/(10/2)*
>
> *LET C8=(C5–50)/(C6/2)*
>
> *CODE (–99:1.638)0 (1.638:99)1 C7–C8 C9–C10*
>
> *TABLE C9*
>
> *TABLE C10*

Note that columns 7 and 8 contain the z- and t-statistics, respectively. A 0 in column 9 indicates that the z-statistic is less than or equal to 1.638, and a 1 indicates that it is greater than 1.638. A 0 in column 10 indicates that the t-statistic is less than or equal to 1.638, and a 1 indicates that it is greater than 1.638. The output counts the number of 0s and 1s in columns 9 and 10.

Instructions for users of full PC versions of Minitab: Type

> *EXECUTE 'A:EXP4–10'*

Instructions for users of student versions of Minitab: Type

> *EXECUTE 'A:EXP4–10S'*

The number of samples to be generated is reduced to 300. We recommend that you repeat this experiment three times and combine the results.

The next experiment will examine the confidence interval estimator of the mean when the population variance is unknown.

EXPERIMENT 4.11

Objective: To illustrate that the 95% confidence interval estimator using the Student t distribution will, in the long run, correctly estimate the mean 95% of the time.

Experiment: Generate 1000 samples of size 4 from a normal population with a mean of 50 and a standard deviation of 10. We will then compute the 95% confidence interval estimates of μ using the Student t distribution and count the number of interval estimates that include the population mean. The 95% confidence interval estimator of the mean using the Student t distribution is

$$\bar{x} \pm t_{\alpha/2} \frac{s}{\sqrt{n}}$$

With $1 - \alpha = .95$, $n = 4$, and d.f. $= 3$ we have

$$\bar{x} \pm t_{\alpha/2} \frac{s}{\sqrt{n}} = \bar{x} \pm 3.182 \frac{s}{\sqrt{4}} = \bar{x} \pm 3.182 \frac{s}{2}$$

Minitab commands:

> *ERASE C1–C49*
>
> *RANDOM 1000 C1–C4;*
>
> *NORMAL 50 10.*
>
> *RMEAN C1–C5*
>
> *RSTDEV C1–C4 C6*
>
> *LET C7=C5–3.182*C6/2*
>
> *LET C8=C5+3.182*C6/2*
>
> *CODE (–99:50)1(50:199)2 C7–C8 C9–C10*
>
> *TABLE C9 C10*

Note that the output of this experiment is to be interpreted in the same way that you interpreted the results of Experiment 4.8.

Instructions for users of full PC versions of Minitab: Type

> *EXECUTE 'A:EXP4–11'*

Instructions for users of student versions of Minitab: Type

> *EXECUTE 'A:EXP4–11S'*

The number of samples to be generated is reduced to 300. We recommend that you repeat this experiment three times and combine the results.

4.6 SELECTING THE SAMPLE SIZE NECESSARY TO ESTIMATE A POPULATION MEAN

In order to determine the sample size to estimate a population mean to within B units with $1 - \alpha$ confidence, we compute n from the formula

$$n = \frac{z_{\alpha/2}^2 \sigma^2}{B^2}$$

Notice that we need a value for σ to find n. This value is usually estimated informally (sometimes colloquially called "guesstimating") from previous knowledge about the population. However, if the the value of σ is underestimated, the confidence interval estimate produced from the sample will be wider than originally planned.

To prepare for the next experiment, find the sample size required to estimate a population mean to within 3 units with 90% confidence. Assume that you believe that the population standard deviation is 10.

EXPERIMENT 4.12

Objective: To illustrate the effect on the confidence interval estimator of using the correct value of σ when determining the sample size.

Experiment: Generate one random sample of the size determined above ($n = 30$) from a normal population with a mean of 50 and a standard deviation of 10, and store the values in column 1. Calculate the 90% confidence interval estimate of μ assuming that σ is unknown.

Minitab commands:

    ERASE C1-C49

    RANDOM 30 C1;

    NORMAL 50 10.

    TINTERVAL .90 C1

Instructions for all users of Minitab: Type

    EXECUTE 'A:EXP4–2'

EXPERIMENT 4.13

Objective: To illustrate the effect on the confidence interval estimate when σ is underestimated when determining the sample size.

Experiment: We repeat Experiment 4.12, generating the sample of size 30 from a normal population whose mean is 50 and whose standard deviation is $\sigma = 20$.

Minitab commands:

    ERASE C1–C49

    RANDOM 30 C1;

    NORMAL 50 20.

    TINTERVAL .90 C1

Instructions for all users of Minitab: Type

    EXECUTE 'A:EXP4–13'

REPORT FOR EXPERIMENT 4.1

If the sample mean is an unbiased estimator of the population mean, what did you anticipate that Minitab would print?

What did Minitab print?

Does it appear reasonable to believe that the sample mean is an unbiased estimator of the population mean? Explain.

REPORT FOR EXPERIMENT 4.2

If the sample median is an unbiased estimator of the population mean, what output did you anticipate that Minitab would print?

What did Minitab print?

Why did we instruct Minitab to print the *mean* of the sample medians?

Does it appear that the sample median is an unbiased estimator of the population mean? Explain.

REPORT FOR EXPERIMENT 4.3

If the sample variance is an unbiased estimator of the population variance, what did you anticipate that Minitab would print?

What did Minitab print?

Does it appear that the sample variance defined as

$$s^2 = \frac{\Sigma(x_i - \bar{x})^2}{n - 1}$$

is an unbiased estimator of the population variance? Explain.

REPORT FOR EXPERIMENT 4.4

If the quantity

$$\frac{\Sigma(x_i - \bar{x})^2}{n}$$

is an unbiased estimator of the population variance, what value did you anticipate Minitab printing?

What did Minitab print?

Does it appear that the quantity

$$\frac{\Sigma(x_i - \bar{x})^2}{n}$$

is an unbiased estimator of the population variance? Explain.

(continued)

Why do you suppose we specified a sample size of only 4 in this experiment?

REPORT FOR EXPERIMENT 4.5

Observed value of the standard deviation of the sampling distribution of the mean with $n = 4$:

REPORT FOR EXPERIMENT 4.6

Observed value of the standard deviation of the sampling distribution of the mean with $n = 25$:

What do the standard deviations computed in Experiments 4.5 and 4.6 tell you about whether $\bar{x}$ is a consistent estimator of μ?

(continued)

Discuss the similarities and differences between the histograms drawn in the two experiments.

What conclusions can you draw from the results of these two experiments?

REPORT FOR EXPERIMENT 4.7

Number of intervals containing the true population mean you anticipated seeing:

Number of intervals containing the true population mean observed in this experiment:

What can you conclude from the results of this experiment?

REPORT FOR EXPERIMENT 4.8

Number of intervals containing the true population mean you anticipated seeing:

Number of intervals containing the true population mean observed in this experiment:

What conclusions can you draw from the results of this experiment?

REPORT FOR EXPERIMENT 4.9

Does it appear that the Student t distribution (with 3 degrees of freedom) is more spread out than the standard normal?

What can you say about their shapes?

What conclusions can you draw from the two *DESCRIBE* commands?

(continued)

Explain why the sampling distribution of

$$\frac{\bar{x} - \mu}{s/\sqrt{n}}$$

is more spread out than the sampling distribution of

$$\frac{\bar{x} - \mu}{\sigma/\sqrt{n}}$$

REPORT FOR EXPERIMENT 4.10

Anticipated number of values of $t > 1.638$:

Observed number of values of $t > 1.638$:

Anticipated number of values of $z > 1.638$:

Observed number of values of $z > 1.638$:

(continued)

What can you conclude from this experiment?

REPORT FOR EXPERIMENT 4.11

Anticipated number of intervals containing the true population mean:

Observed number of intervals containing the true population mean:

What conclusion can you draw from the results of this experiment?

REPORT FOR EXPERIMENT 4.12

Size of the interval that we wanted:

Size of the actual interval calculated in this experiment:

REPORT FOR EXPERIMENT 4.13

Size of the interval that we wanted:

Size of the actual interval calculated in this experiment:

(*continued*)

What conclusion should be drawn from Experiments 4.12 and 4.13 when guesstimating σ to determine the sample size needed to estimate μ?

CHAPTER 5

Hypothesis Testing

5.1 INTRODUCTION

Hypothesis testing is another form of statistical inference. Whereas estimation presumes no knowledge of the value of a parameter, hypothesis testing assumes that you have some value of a parameter that you wish to test. A test of hypothesis consists of four components:

1. Null hypothesis
2. Alternative hypothesis
3. Test statistic
4. Rejection region

Null Hypothesis

The null hypothesis, which is denoted H_0 (H-naught), specifies a single value for the population parameter. For example, if we want to test to determine whether the mean weight-loss of people who have participated in a new weight program is 10 pounds, we would test

$$H_0: \mu = 10$$

Alternative Hypothesis

This hypothesis, denoted H_A is the more important one because it is the hypothesis that answers our question. The alternative hypothesis can assume three possible forms:

1. If a tire company wanted to know whether the average life of its new radial tire exceeds its advertised value of 50,000 miles, the company would specify the alternative hypothesis as

    $$H_A: \mu > 50,000$$

2. If the company wanted to know whether the average life of the tire is less than 50,000 miles, it would test

    $$H_A: \mu < 50,000$$

3.   If the company wished to determine whether the average life of the tire differs from that advertised, its alternative hypothesis would be

$$H_A: \mu \neq 50{,}000$$

In all three cases the null hypothesis would be

$$H_0: \mu = 50{,}000$$

Test Statistic

The test statistic is the sample statistic upon which we base our decision to either reject or not reject the null hypothesis. For parametric tests of hypotheses, the test statistic is the standardized point estimator of the parameter being tested. For example, since the sample mean $\bar{x}$ is the point estimator of the population mean μ, the test statistic for hypotheses about μ (assuming σ is known) is

$$z = \frac{\bar{x} - \mu}{\sigma/\sqrt{n}}$$

Rejection Region

The rejection region is a range of values such that, if the test statistic falls into that range, we decide to reject the null hypothesis. To illustrate, suppose that we wish to test

$$H_0: \mu = 1000$$

If we find that the sample mean is quite different from 1000, and thus that z is quite different from 0, we say that the test statistic falls into the rejection region and we reject the null hypothesis. On the other hand, if $\bar{x}$ is close to 1000, (and z is close to 0) we cannot reject the null hypothesis. The key question answered by the rejection region is, When is the value of the test statistic sufficiently different from the hypothesized value of the parameter to enable us to reject the null hypothesis? The process we use in answering this question depends on the probability of our making a mistake when testing the hypothesis.

Since the conclusion we draw is based on sample data, the chance of our making one of two possible errors will always exist. As indicated in Figure 5.1, the null hypothesis is either true or false, and we must decide either to reject it or not reject it. Therefore, two correct decisions are possible: rejecting the null hypothesis when it is false, and not rejecting the null hypothesis when it is true. Conversely, two incorrect decisions are possible: rejecting H_0 when it is true (this is called a Type I error, and the probability of committing it is α), and not rejecting H_0 when it is false (this is called a Type II error, and the probability of committing it is β). The probability α is called the significance level.

FIGURE 5.1 RESULTS OF A TEST OF HYPOTHESIS

	H_0 is true	H_0 is false
Do not reject H_0	Correct decision	Type II error P(Type II error) = β
Reject H_0	Type I error P(Type I error) = α	Correct decision

We would like both α and β to be as small as possible, but unfortunately there is an inverse relationship between α and β. Thus, for a given sample size, any decrease in α results in an increase in β.

The goal of the experiments described in this chapter is to help you understand Type I and Type II errors.

5.2 TYPE I ERRORS

EXPERIMENT 5.1

Objective: To illustrate that, when testing a true null hypothesis, Type I errors will be committed at a rate equal to the significance level.

Experiment: We will generate 1000 samples of size 4 from a normal population whose mean and standard deviation are 100 and 20, respectively. For each sample, we will conduct the following test of hypotheses:

H_0: μ = 100

H_A: $\mu \neq 100$

The test will be conducted with a significance level of 5%, and we will assume that $\sigma = 20$. The test statistic and the rejection region are

Test statistic: $z = \dfrac{\bar{x} - \mu}{\sigma/\sqrt{n}}$

Rejection region: $|z| > z_{\alpha/2} = z_{.025} = 1.96$

Since the null hypothesis is true, σ is equal to 20, and n is equal to 4, the value of the test statistic is

$$z = \frac{\bar{x} - \mu}{\sigma/\sqrt{n}} = \frac{\bar{x} - 100}{20/\sqrt{4}} = \frac{\bar{x} - 100}{10}$$

Minitab commands:

ERASE C1–C49

RANDOM 1000 C1–C4;

NORMAL 100 20.

RMEAN C1–C5

LET C6=(C5–100)/10

CODE (–99:–1.96)1 (–1.96:1.96)0 (1.96:99)1 C6 C7

TABLE C7

Note that column 6 stores the values of the test statistics and that column 7 contains the coded values. A 1 in column 7 represents a test statistic that is in the rejection region ($z < -1.96$ or $z > 1.96$). Since the null hypothesis is true, a 1 represents a Type I error. A 0 in column 7 indicates a test statistic that does not fall into the rejection region. Minitab outputs the number of 0s and 1s in column 7.

Instructions for users of full PC versions of Minitab: Type

EXECUTE 'A:EXP5–1'

Instructions for users of student versions of Minitab: Type

EXECUTE 'A:EXP5–1S'

The number of samples to be generated is reduced to 500. We recommend that you repeat this experiment twice and combine the results.

EXPERIMENT 5.2

Objective: Together with Experiment 5.1, to illustrate Type I errors.

Experiment: Repeat Experiment 5.1 using a significance level of 10%. The rejection region is

$$|z| > z_{\alpha/2} = z_{.05} = 1.645$$

Minitab commands:

ERASE C1–C49

RANDOM 1000 C1–C4;

NORMAL 100 20.

RMEAN C1–C5

LET C6=(C5–100)/10

CODE (–99:–1.645)1(–1.645:1.645)0(1.645:99)1 C6 C7

TABLE C7

Instructions for users of full PC versions of Minitab: Type

EXECUTE 'A:EXP5–2'

Instructions for users of student versions of Minitab: Type

EXECUTE 'A:EXP5–2S'

The number of samples to be generated is reduced to 500. We recommend that you repeat this experiment twice and combine the results.

5.3   TYPE II ERRORS

As we discussed, it is possible to make two types of error when conducting a test of hypothesis. A Type II error occurs when we do not reject a false null hypothesis. Consider the example discussed in Experiment 5.1. The null and alternative hypotheses were

$$H_0: \mu = 100$$

$$H_A: \mu \neq 100$$

The test was conducted at the 5% significance level, assuming that $\sigma = 20$, and $n = 4$. The rejection region was

$$|z| > 1.96$$

The next six experiments address the issue of calculating the probability of a Type II error under different conditions.

Suppose that we want to compute the probability of a Type II error when μ is actually equal to 110.

EXPERIMENT 5.3

Objective: To illustrate the occurrence of Type II errors and, together with the next two experiments, to illustrate the effect of the sample size on the probability of Type II errors.

Experiment: We will generate 1000 samples of size 4 from a normal population with a mean of 110 and a standard deviation of 20. Using these data, we test the following hypotheses at the 5% significance level.

$$H_0: \mu = 100$$

$$H_A: \mu \neq 100$$

The test statistic is

$$z = \frac{\bar{x} - \mu}{\sigma/\sqrt{n}} = \frac{\bar{x} - 100}{20/\sqrt{4}} = \frac{\bar{x} - 100}{10}$$

and the rejection region is

$$|z| > 1.96$$

Since the null hypothesis is false (μ is really equal to 110), any values of the test statistic that do not lead to the rejection of the null hypothesis will result in Type II errors.

Minitab commands:

```
ERASE C1–C49

RANDOM 1000 C1–C4;

NORMAL 110 20.
```

RMEAN C1–C5

LET C6=(C5–100)/10

CODE (–99:–1.96)1(–1.96:1.96)0(1.96:99)1 C6 C7

TABLE C7

Instructions for users of full PC versions of Minitab: Type

EXECUTE 'A:EXP5–3'

Instructions for users of student versions of Minitab: Type

EXECUTE 'A:EXP5–3S'

The number of samples to be generated is reduced to 500. We recommend that you repeat this experiment twice and combine the results.

EXPERIMENT 5.4

Objective: In concert with experiment 5.3 and 5.5, to illustrate the effect of increasing the sample size on the probability of committing a Type II error.

Experiment: We repeat Experiment 5.3 using a sample size of 9. The value of the test statistic is computed as follows:

$$z = \frac{\bar{x} - \mu}{\sigma/\sqrt{n}} = \frac{\bar{x} - 100}{20/\sqrt{9}} = \frac{\bar{x} - 100}{6.667}$$

Minitab commands:

ERASE C1–C49

RANDOM 1000 C1–C9;

NORMAL 110 20.

RMEAN C1–C10

LET C11=(C10–100)/6.667

CODE (–99:–1.96)1(–1.96:1.96)0(1.96:99)1 C11 C12

TABLE C12

Instructions for users of full PC versions of Minitab: Type

   EXECUTE 'A:EXP5–4'

Instructions for users of student versions of Minitab: Type

   EXECUTE 'A:EXP5–4S'

The number of samples to be generated is reduced to 200. We recommend that you rerun this experiment five times and combine results.

EXPERIMENT 5.5

Objective: Together with Experiments 5.3 and 5.4, to illustrate the effect of increasing the sample size on the probability of a Type II error.

Experiment: We will repeat Experiment 5.3 using a sample size of 25. We reduce the number of samples to complete this experiment. The value of the test statistic is

$$z = \frac{\bar{x} - \mu}{\sigma/\sqrt{n}} = \frac{\bar{x} - 100}{20/\sqrt{25}} = \frac{\bar{x} - 100}{4}$$

Minitab commands:

   ERASE C1–C49

   RANDOM 500 C1–C25;

   NORMAL 110 20.

   RMEAN C1–C26

   LET C27=(C26–100)/4

   CODE (–99:–1.96)1(–1.96:1.96)0(1.96:99)1 C27 C28

   TABLE C28

Instructions for users of full PC versions of Minitab: Type

   EXECUTE 'A:EXP5–5'

Instructions for users of student versions of Minitab: Type

   EXECUTE 'A:EXP5–5S'

The number of samples to be generated is reduced to 100. Run this experiment five times and combine results.

EXPERIMENT 5.6

Objective: In concert with Experiments 5.3 and 5.7, to illustrate the effect of changing the significance level on the probability of a Type II error.

Experiment: We will repeat Experiment 5.3 using $\alpha = .10$. The rejection region is

$$|z| > z_{\alpha/2} = z_{.05} = 1.645$$

Minitab commands:

        ERASE C1–C49

        RANDOM 1000 C1–C4;

        NORMAL 110 20.

        RMEAN C1–C5

        LET C6=(C5–100)/10

        CODE (–99:–1.645)1(–1.645:1.645)0(1.645:99)1 C6 C7

        TABLE C7

Instructions for users of full PC versions of Minitab: Type

        EXECUTE 'A:EXP5-6'

Instructions for users of student versions of Minitab: Type

        EXECUTE 'A:EXP5-6S'

The number of samples to be generated is reduced to 500. We recommend that you repeat this experiment twice and combine the results.

EXPERIMENT 5.7

Objective: Together with Experiments 5.3 and 5.6, to illustrate the effect of changing the significance level on the probability of a Type II error.

Experiment: We will repeat Experiment 5.3 using a significance level of 1%. The rejection region is

$$|z| > z_{\alpha/2} = z_{.005} = 2.575$$

Minitab commands:

> ERASE C1–C49
>
> RANDOM 1000 C1–C4;
>
> NORMAL 110 20.
>
> RMEAN C1–C5
>
> LET C6 = (C5–100)/10
>
> CODE (–99:–2.575)1(–2.575:2.575)0(2.575:99)1 C6 C7
>
> TABLE C7

Instructions for users of full PC versions of Minitab: Type

> EXECUTE 'A:EXP5–7'

Instructions for users of student versions of Minitab: Type

> EXECUTE 'A:EXP5–7S'

The number of samples to be generated is reduced to 500. We recommend that you repeat this experiment twice and combine the results.

EXPERIMENT 5.8

Objective: To illustrate the effect of changing the actual value of μ on the probability of a Type II error.

Experiment: We repeat Experiment 5.3 using the following values of μ: 60, 80, 100, 120, 140

Minitab commands:

> ERASE C1–C49
>
> RANDOM 1000 C1–C4;

NORMAL 60 20.

RMEAN C1-C5

LET C6=(C5-100)/10

CODE (-99:-1.96)1(-1.96:1.96)0(1.96:99)1 C6 C7

TABLE C7

ERASE C1-C49

RANDOM 1000 C1-C4;

NORMAL 80 20.

RMEAN C1-C5

LET C6=(C5-100)/10

CODE (-99:-1.96)1(-1.96:1.96)0(1.96:99)1 C6 C7

TABLE C7

ERASE C1-C49

RANDOM 1000 C1-C4;

NORMAL 100 20.

RMEAN C1-C5

LET C6=(C5-100)/10

CODE (-99:-1.96)1(-1.96:1.96)0(1.96:99)1 C6 C7

TABLE C7

ERASE C1-C49

RANDOM 1000 C1-C4;

NORMAL 120 20.

RMEAN C1-C5

LET C6=(C5–100)/10

CODE (–99:–1.96)1(–1.96:1.96)0(1.96:99)1 C6 C7

TABLE C7

ERASE C1–C49

RANDOM 1000 C1–C4;

NORMAL 140 20.

RMEAN C1–C5

LET C6=(C5–100)/10

CODE (–99:–1.96)1(–1.96:1.96)0(1.96:99)1 C6 C7

TABLE C7

Note that there will be five tables output, each representing the results for a different value of μ.

Instructions for users of full PC versions of Minitab: Type

EXECUTE 'A:EXP5–8'

Instructions for users of student versions of Minitab: Type

EXECUTE 'A:EXP5–8S'

The number of samples to be generated is reduced to 500.

REPORT FOR EXPERIMENT 5.1

Anticipated percentage of Type I errors:

Observed percentage of Type I errors:

REPORT FOR EXPERIMENT 5.2

Anticipated percentage of Type I errors:

Observed percentage of Type I errors:

(*continued*)

Explain why, when a Type I error is committed, it is not necessarily because of some mistake the statistician made:

Discuss the implications of Experiments 5.1 and 5.2.

REPORT FOR EXPERIMENT 5.3

Observed percentage of Type II errors:

REPORT FOR EXPERIMENT 5.4

Observed percentage of Type II errors:

REPORT FOR EXPERIMENT 5.5

Observed percentage of Type II errors:

(*continued*)

Compare the results of Experiments 5.3, 5.4, and 5.5. What effect does increasing the sample size have on the probability of a Type II error?

REPORT FOR EXPERIMENT 5.6

Observed percentage of Type II errors:

REPORT FOR EXPERIMENT 5.7

Observed percentage of Type II errors:

Compare the results of Experiments 5.3, 5.6, and 5.7. What effect does the significance level have on the probability of a Type II error?

REPORT FOR EXPERIMENT 5.8

$\mu = 60$: Observed percentage of Type II errors:

$\mu = 80$: Observed percentage of Type II errors:

$\mu = 100$: Observed percentage of Type II errors:

$\mu = 120$: Observed percentage of Type II errors:

$\mu = 140$: Observed percentage of Type II errors:

Discuss what effect the value of μ has on the probability of making a Type II error.

CHAPTER 6

Violations of Required Conditions

6.1 INTRODUCTION

All statistical procedures have a number of required conditions that have to be satisfied to ensure that the results of the procedures are valid. In this chapter we will examine what happens when the required conditions are not satisfied.

Note that because of the problem of drawing conclusions based on a relatively small number of samples in this chapter, we will recommend that each experiment be repeated several times. Specific instructions are provided with each experiment.

6.2 NORMALITY REQUIREMENT

Many statistical techniques require that the population being sampled be normal. These include the t-test of μ, the t-test of $(\mu_1 - \mu_2)$, and the F-test of σ_1^2/σ_2^2.

In this section we'll describe several experiments that allow you to discover what happens when the normality requirement is not satisfied. Experiments 6.1 to 6.7 compare the percentage of Type I errors committed by the t-test of μ when the populations are normal, uniform and exponential, and for different sample sizes. Experiments 6.8 and 6.9 compare the percentage of Type II errors.

EXPERIMENT 6.1

Objective: To illustrate that when the population is normal, the percentage of Type I errors committed by the t-test of μ is equal to the significance level.

Experiment: We will generate 1000 samples of size 4 from a normal population whose mean is 10 and whose standard deviation is 5. For each sample, we will test the hypotheses

$$H_0: \mu = 10$$

$$H_A: \mu \neq 10$$

We assume that the population is normal. The significance level is set at 10%. The test statistic is

$$t = \frac{\bar{x} - \mu}{s/\sqrt{n}}$$

and the rejection region is

$$|t| > t_{\alpha/2,\, n-1} = t_{.05,\, 3} = 2.353$$

We will count the number of times a Type I error is committed.

Minitab commands:

    ERASE C1–C49

    RANDOM 1000 C1–C4;

    NORMAL 10 5.

    RMEAN C1–C5

    RSTDEV C1–C4 C6

    LET C7=(C5–10)/(C6/2)

    CODE (–99:–2.353)–1(–2.353:2.353)0(2.353:99)1 C7 C8

    TABLE C8

The values of the test statistic are stored in column 7. Column 8 contains coded values such that, when the test statistic does not fall into the rejection region, the value is 0. Values of $t < -2.353$ are coded –1 and values of $t > 2.353$ are coded 1.

Instructions for users of full PC versions of Minitab: Type

    EXECUTE 'A:EXP6–1'

We recommend that you repeat this experiment twice and combine the results.

Instructions for users of student versions of Minitab: Type

    EXECUTE 'A:EXP6–1S'

The number of samples to be generated is reduced to 400. We recommend that you repeat this experiment five times and combine the results.

EXPERIMENT 6.2

Objective: To illustrate what happens to the probability of a Type I error in the *t*-test of μ when the population is moderately nonnormal.

Experiment: We repeat Experiment 6.1, generating 1000 samples of size 4 from a uniform distribution with a minimum value of 0 and a maximum value of 20. (The mean of this distribution is 10.)

Minitab commands:

ERASE C1–C49

RANDOM 1000 C1–C4;

UNIFORM 0 20.

RMEAN C1–C5

RSTDEV C1–C4 C6

LET C7=(C5–10)/(C6/2)

CODE (–99:–2.353)–1(–2.353:2.353)0(2.353:99)1 C7 C8

TABLE C8

Instructions for users of full PC versions of Minitab: Type

EXECUTE 'A:EXP6–2'

We recommend that you repeat this experiment twice and combine the results.

Instructions for users of student versions of Minitab: Type

EXECUTE 'A:EXP6–2S'

The number of samples to be generated is reduced to 400. We recommend that you repeat this experiment five times and combine the results.

EXPERIMENT 6.3

Objective: In concert with Experiments 6.2 and 6.4, to examine the effect of increasing the sample size on the probability of a Type I error in the *t*-test of μ when the population is moderately nonnormal.

Experiment: We will repeat Experiment 6.2 using a sample of size 9.

Minitab commands:

ERASE C1–C49

RANDOM 1000 C1–C9;

UNIFORM 0 20.

RMEAN C1–C10

RSTDEV C1–C9 C11

LET C12=(C10–10)/(C11/3)

CODE (–99:–1.860)–1(–1.860:1.860)0(1.860:99)1 C12 C13

TABLE C13

Notice that, because the degrees of freedom are 8, the rejection region is now

$$|t| > 1.860$$

Instructions for users of full PC versions of Minitab: Type

EXECUTE 'A:EXP6–3'

We recommend that you repeat this experiment twice and combine the results.

Instructions for users of student versions of Minitab: Type

EXECUTE 'A:EXP6–3S'

The number of samples to be generated is reduced to 200. We recommend that you repeat this experiment ten times and combine results.

EXPERIMENT 6.4

Objective: In concert with Experiments 6.2 and 6.3, to illustrate the effect of increasing the sample size on the probability of a Type I error in the t-test of μ when the population is moderately nonnormal.

Experiment: We repeat Experiment 6.2 using $n = 25$. We reduce the number of samples to 500.

Minitab commands:

    ERASE C1–C49

    RANDOM 500 C1–C25;

    UNIFORM 0 20.

    RMEAN C1–C26

    RSTDEV C1–C25 C27

    LET C28=(C26–10)/(C27/5)

    CODE (–99:–1.711)–1(–1.711:1.711)0(1.711:99)1 C28 C29

    TABLE C29

Note that the rejection region is

    $|t| > 1.711$

Instructions for users of full PC versions of Minitab: Type

    EXECUTE 'A:EXP6–4'

We recommend that you repeat this experiment twice and combine the results.

Instructions for users of student versions of Minitab: Type

    EXECUTE 'A:EXP6–4S'

The number of samples to be generated is reduced to 100. We recommend that you repeat this experiment ten times and combine the results.

EXPERIMENT 6.5

Objective: To illustrate what happens to the probability of a Type I error in the *t*-test of μ when the population is extremely nonnormal.

Experiment: We repeat Experiment 6.1, assuming that the population is exponential with a mean of 10.

Minitab commands:

> *ERASE C1–C49*
>
> *RANDOM 1000 C1–C4;*
>
> *EXPONENTIAL 10.*
>
> *RMEAN C1–C5*
>
> *RSTDEV C1–C4 C6*
>
> *LET C7=(C5–10)/(C6/2)*
>
> *CODE (–99:–2.353)–1(–2.353:2.353)0(2.353:99)1 C7 C8*
>
> *TABLE C8*

Instructions for users of full PC versions of Minitab: Type

> *EXECUTE 'A:EXP6–5'*

We recommend that you repeat this experiment twice and combine the results.

Instructions for users of student versions of Minitab: Type

> *EXECUTE 'A:EXP6–5S'*

The number of samples to be generated is reduced to 400. We recommend that you repeat this experiment five times and combine the results.

EXPERIMENT 6.6

Objective: Together with Experiments 6.5 and 6.7, to illustrate what happens to the probability of a Type I error in the t-test of μ when we increase the sample size and the population is extremely nonnormal.

Experiment: We repeat Experiment 6.5 using $n = 9$.

Minitab commands:

> *ERASE C1–C49*
>
> *RANDOM 1000 C1–C9;*

EXPONENTIAL 10.

RMEAN C1–C10

RSTDEV C1–C9 C11

LET C12=(C10–10)/(C11/3)

CODE (–99:–1.860)–1(–1.860:1.860)0(1.860:99)1 C12 C13

TABLE C13

Instructions for users of full PC versions of Minitab: Type

EXECUTE 'A:EXP6-6'

We recommend that you repeat this experiment twice and combine the results.

Instructions for users of student versions of Minitab: Type

EXECUTE 'A:EXP6-6S'

The number of samples to be generated is reduced to 200. We recommend that this experiment be repeated ten times and the results combined.

EXPERIMENT 6.7

Objective: Together with Experiments 6.5 and 6.6, to illustrate the effect of increasing the sample size on the probability of a Type I error in the test of μ when the population is extremely nonnormal.

Experiment: We repeat Experiment 6.5 using $n = 25$.

Minitab commands:

ERASE C1–C49

RANDOM 500 C1–C25;

EXPONENTIAL 10.

RMEAN C1–C26

RSTDEV C1–C25 C27

LET C28=(C26-10)/(C27/5)

CODE (-99:-1.711)-1(-1.711:1.711)0(1.711:99)1 C28 C29

TABLE C29

Instructions for users of full PC versions of Minitab: Type

EXECUTE 'A:EXP6-7'

We recommend that you repeat this experiment twice and combine the results.

Instructions for users of student versions of Minitab: Type

EXECUTE 'A:EXP6-7S'

The number of samples to be generated is reduced to 100. This experiment should be repeated ten times and the results combined.

The next two experiments examine the effects of nonnormality on the probability of Type II errors.

EXPERIMENT 6.8

Objective: In concert with Experiment 6.9, to illustrate that when the population is normal, we commit a smaller percentage of Type II errors than we do when the population is moderately nonnormal.

Experiment: We will generate 500 samples of size 4 from each of two populations. Population 1 is normal with mean 10 and standard deviation 5.77, and population 2 is uniform ranging from 0 to 20. The mean and standard deviation of the normal population were selected so that the two populations have equal means and standard deviations. For each population, we compute the test statistics to test the hypotheses

H_0: μ = 5

H_A: $\mu \neq 5$

With α = .05, the rejection region is

$$|t| = t_{\alpha/2, n-1} = t_{.025, 3} = 3.182$$

Since the null hypothesis is false (the actual population means are 10), any samples that do not reject the null hypothesis will result in Type II errors.

Minitab commands:

> *ERASE C1–C49*
>
> *RANDOM 500 C1–C4;*
>
> *NORMAL 10 5.77.*
>
> *RANDOM 500 C5–C8;*
>
> *UNIFORM 0 20.*
>
> *RMEAN C1–C4 C9*
>
> *RSTDEV C1–C4 C10*
>
> *RMEAN C5–C8 C11*
>
> *RSTDEV C5–C8 C12*
>
> *LET C13=(C9–5)/(C10/2)*
>
> *LET C14=(C11–5)/(C12/2)*
>
> *CODE (–99:–3.182)1(–3.182:3.182)0(3.182:99)1 C13–C14 C15–C16*
>
> *TABLE C15*
>
> *TABLE C16*

Columns 1 to 4 contain the samples from the normal population, and columns 5 to 8 store the data from the uniform population. Columns 9 through 12 contain the sample means and standard deviations. In columns 13 and 14, the test statistics are stored. The number of times the null hypothesis is rejected (code 1) and the number of times it is not rejected (code 0) are printed. The results of sampling from the normal population appears first, followed by the results of sampling from the uniform population.

Instructions for users of full PC versions of Minitab: Type

> *EXECUTE 'A:EXP6–8'*

We recommend that you repeat this experiment twice and combine the results.

Instructions for users of student versions of Minitab: Type

    EXECUTE 'A:EXP6–8S'

The number of samples to be generated is reduced to 200. We recommend that you repeat this experiment ten times and combine the results.

EXPERIMENT 6.9

Objective: Together with Experiment 6.8, to illustrate what happens to the probability of Type II errors from normal and moderately nonnormal populations when the sample size is increased.

Experiment: We repeat Experiment 6.8, increasing the sample size to 16.

Minitab commands:

    ERASE C1–C49

    RANDOM 250 C1–C16;

    NORMAL 10 5.77.

    RANDOM 250 C17–C32;

    UNIFORM 0 20.

    RMEAN C1–C16 C33

    RSTDEV C1–C16 C34

    RMEAN C17–C32 C35

    RSTDEV C17–C32 C36

    LET C37=(C33–5)/(C34/4)

    LET C38=(C35–5)/(C36/4)

    CODE (–99:–2.131)1(–2.131:2.131)0(2.131:99)1 C37–C38 C39–C40

    TABLE C39

    TABLE C40

Note that we reduced the number of samples to 250, and the rejection region is

$$|t| > t_{\alpha/2,\, n-1} = t_{.025,\, 15} = 2.131$$

Instructions for users of full PC versions of Minitab: Type

EXECUTE 'A:EXP6-9'

We recommend that you repeat this experiment twice and combine the results.

Instructions for users of student versions of Minitab: Type

EXECUTE 'A:EXP6-9S'

The number of samples to be generated is reduced to 50. We recommend that you repeat this experiment ten times and combine the results.

6.3 EQUAL VARIANCES REQUIREMENT

As well as requiring that the populations be normal, many statistical techniques that compare two or more populations of quantitative data require that the population variances be equal. One such test is the t-test of $(\mu_1 - \mu_2)$. When the sample sizes are small (less than 30), the validity of the test results requires that $\sigma_1^2 = \sigma_2^2$. The next two experiments examine this requirement.

Suppose that we want to test the hypotheses

$$H_0: (\mu_1 - \mu_2) = 0$$

$$H_A: (\mu_1 - \mu_2) \neq 0$$

at the 10% significance level, using samples of size 4. The test statistic is

$$t = \frac{(\bar{x}_1 - \bar{x}_2) - (\mu_1 - \mu_2)}{\sqrt{s_p^2 \left(\dfrac{1}{n_1} + \dfrac{1}{n_2} \right)}} \qquad \text{d.f.} = n_1 + n_2 - 2 = 6$$

and the rejection region is

$$|t| > t_{\alpha/2,\, n_1 + n_2 - 2} = t_{.05,\, 6} = 1.943$$

EXPERIMENT 6.10

Objective: To illustrate that when the normality and equal variances requirements are satisfied, the percentage of Type I errors in the t-test of $(\mu_1 - \mu_2)$ is equal to the significance level.

Experiment: Minitab will generate 1000 pairs of samples from a normal population with a mean of 50 and a standard deviation of 10. For each pair of samples, we will calculate the test statistic and count the number of Type I errors.

Minitab commands:

```
ERASE C1-C49

RANDOM 1000 C1-C8;

NORMAL 50 10.

RMEAN C1-C4 C9

RSTDEV C1-C4 C10

RMEAN C5-C8 C11

RSTDEV C5-C8 C12

LET C13=(C10**2+C12**2)/2

LET C14=(C9-C11)/((C13/2)**0.5)

CODE (-99:-1.943)-1(-1.943:1.943)0(1.943:99)1 C14 C15

TABLE C15
```

We treat the numbers in columns 1 to 4 as the samples of size 4 from population 1. The numbers in columns 5 to 8 represent samples of size 4 from population 2. Both populations have equal means ($\mu_1 = \mu_2 = 50$) and equal variances ($\sigma_1^2 = \sigma_2^2 = 10^2 = 100$). We compute the sample means and standard deviations (columns 9 to 12). From the standard deviations, we calculate the pooled variance estimate, s_p^2, which is stored in column 13. Column 14 stores the values of the test statistics, and column 15 contains the codes indicating whether the test statistic is in the rejection region (code -1 or $+1$) or not (code 0).

Instructions for users of full PC versions of Minitab: Type

    EXECUTE 'A:EXP6–10'

Instructions for users of student versions of Minitab: Type

    EXECUTE 'A:EXP6–10S'

The number of samples to be generated is reduced to 200. We recommend that you repeat this experiment five times and combine the results.

EXPERIMENT 6.11

Objective: In concert with Experiment 6.10, to illustrate that when the equal variances requirement in the t-test of $(\mu_1 - \mu_2)$ is not satisfied, the percentage of Type I errors is not equal to the significance level.

Experiment: We repeat Experiment 6.10, letting $\sigma_1 = 10$ and $\sigma_2 = 200$.

Minitab commands:

    ERASE C1–C49

    RANDOM 1000 C1–C4;

    NORMAL 50 10.

    RANDOM 1000 C5–C8;

    NORMAL 50 200.

    RMEAN C1–C4 C9

    RSTDEV C1–C4 C10

    RMEAN C5–C8 C11

    RSTDEV C5–C8 C12

    LET C13=(C10**2+C12**2)/2

    LET C14=(C9–C11)/((C13/2)**0.5)

CODE (-99:-1.943)-1(-1.943:1.943)0(1.943:99)1 C14 C15

TABLE C15

The numbers in columns 1 to 4 (sample 1) are drawn from a normal population with a mean of 50 and a standard deviation of 10. Columns 5 through 8 (sample 2) contain numbers that were drawn from a normal population with a mean of 50 and a standard deviation of 200. Thus, the means are equal (i.e., the null hypothesis is true), but the required condition for the t-test of $(\mu_1 - \mu_2)$ is not satisfied.

Instructions for users of full PC versions of Minitab: Type

EXECUTE 'A:EXP6-11'

Instructions for users of student versions of Minitab: Type

EXECUTE 'A:EXP6-11S'

The number of samples to be generated is reduced to 200. We recommend that you repeat this experiment five times and combine the results.

6.4   EFFECT OF NONNORMALITY ON OTHER TESTS

In performing the experiments described in Section 6.2, you discovered that the conclusions drawn from the t-test of μ are affected by the distribution of the population. The question arises, Are all statistical inference procedures equally affected by nonnormality? In this section we present seven experiments designed to compare the effects of nonnormality on the t-test of $\mu_1 - \mu_2$ and the F-test of σ_1^2/σ_2^2.

Suppose that we want to test the following hypotheses:

$$H_0: (\mu_1 - \mu_2) = 0$$

$$H_A: (\mu_1 - \mu_2) \neq 0$$

using a significance level of 10%, with sample sizes of $n_1 = n_2 = 4$. The test statistic is

$$t = \frac{(\bar{x}_1 - \bar{x}_2) - (\mu_1 - \mu_2)}{\sqrt{s_p^2\left(\frac{1}{n_1} + \frac{1}{n_2}\right)}} \quad \text{d.f.} = n_1 + n_2 - 2 = 6$$

and the rejection region is

$$|t| > t_{\alpha/2,\, n_1 + n_2 - 2} = t_{.05,\, 6} = 1.943$$

Note that Experiment 6.10 illustrated the results of this test when the required conditions (normality and equal variances) were satisfied.

EXPERIMENT 6.12

Objective: In concert with Experiment 6.10, to illustrate that when the normality requirement is unsatisfied in the t-test of $\mu_1 - \mu_2$, the percentage of Type I errors is not equal to the significance level.

Experiment: We repeat Experiment 6.10, generating 1000 pairs of samples from an exponential population whose means (and standard deviations) are 50.

Minitab commands:

ERASE C1–C49

RANDOM 1000 C1–C8;

EXPONENTIAL 50.

RMEAN C1–C4 C9

RSTDEV C1–C4 C10

RMEAN C5–C8 C11

RSTDEV C5–C8 C12

LET C13=(C10**2+C12**2)/2

LET C14=(C9–C11)/((C13/2)**0.5)

CODE (–99:–1.943)–1(–1.943:1.943)0(1.943:99)1 C14 C15

TABLE C15

As in Experiment 6.10, the null hypothesis is true. Our objective is to count the number of Type I errors, coded –1 and 1 in the printout.

Instructions for users of full PC versions of Minitab: Type

> *EXECUTE 'A:EXP6–12'*

Instructions for users of student versions of Minitab: Type

> *EXECUTE 'A:EXP6–12S'*

The number of samples to be generated is reduced to 200. We recommend that you repeat this experiment five times and combine the results.

Consider the following hypotheses.

$$H_0: (\mu_1 - \mu_2) = 10$$

$$H_A: (\mu_1 - \mu_2) \neq 10$$

The test statistic and rejection region are the same as specified in Experiments 6.10 and 6.12. The next two experiments generate random samples from populations with equal means. Thus, any tests that do not reject the null hypothesis will commit Type II errors.

EXPERIMENT 6.13

Objective: Together with Experiment 6.14, to illustrate the effect of nonnormality on the probability of a Type II error in the t-test of $\mu_1 - \mu_2$. In this experiment we estimate the probability of a Type II error when the populations are normal.

Experiment: We repeat Experiment 6.10, except we test $H_0: (\mu_1 - \mu_2) = 10$.

Minitab commands:

> *ERASE C1–C49*
>
> *RANDOM 1000 C1–C8;*
>
> *NORMAL 50 10.*
>
> *RMEAN C1–C4 C9*
>
> *RSTDEV C1–C4 C10*
>
> *RMEAN C5–C8 C11*
>
> *RSTDEV C5–C8 C12*

LET C13=(C10**2+C12**2)/2

LET C14=(C9−C11−10)/((C13/2)**0.5)

CODE (−99:−1.943)−1(−1.943:1.943)0(1.943:99)1 C14 C15

TABLE C15

As before, column 14 contains the values of the test statistics. Note, however, in this experiment the value of the test statistic is

$$t = \frac{(\bar{x}_1 - \bar{x}_2) - (10)}{\sqrt{s_p^2 \left(\dfrac{1}{n_1} + \dfrac{1}{n_2} \right)}}$$

Because the null hypothesis is false ($\mu_1 = \mu_2 = 50$), the objective is fulfilled by counting the number of Type II errors, which is indicated by the number of 0s in the printout.

Instructions for users of full PC versions of Minitab: Type

EXECUTE 'A:EXP6−13'

Instructions for users of student versions of Minitab: Type

EXECUTE 'A:EXP6−13S'

The number of samples to be generated is reduced to 200. We recommend that you repeat this experiment five times and combine the results.

EXPERIMENT 6.14

Objective: Together with Experiment 6.13, to illustrate the effect of nonnormality on the probability of Type II errors in the *t*-test of $\mu_1 - \mu_2$.

Experiment: We repeat Experiment 6.13 with exponential populations (with equal means).

Minitab commands:

ERASE C1−C49

RANDOM 1000 C1−C8;

EXPONENTIAL 50.

RMEAN C1–C4 C9

RSTDEV C1–C4 C10

RMEAN C5–C8 C11

RSTDEV C5–C8 C12

LET C13=(C10**2+C12**2)/2

LET C14=(C9–C11–10)/((C13/2)**0.5)

CODE (–99:–1.943)–1(–1.943:1.943)0(1.943:99)1 C14 C15

TABLE C15

Instructions for users of full PC versions of Minitab: Type

EXECUTE 'A:EXP6–14'

Instructions for users of student versions of Minitab: Type

EXECUTE 'A:EXP6–14S'

The number of samples to be generated is reduced to 200. We recommend that you repeat this experiment five times and combine the results.

In Experiments 6.10 and 6.12—6.14, we examined the probability of Type I and Type II errors in the t-test of $\mu_1 - \mu_2$ when the populations are normal and exponential. We now repeat these experiments, but we test the equality of the population variances. That is, we test the hypotheses

$$H_0: \sigma_1^2/\sigma_2^2 = 1$$

$$H_A: \sigma_1^2/\sigma_2^2 \neq 1$$

The test statistic is

$$F = s_1^2/s_2^2 \text{ with degrees of freedom: } v_1 = n_1 - 1 \text{ and } v_2 = n_2 - 1.$$

With samples of size $n_1 = n_2 = 4$ and a significance level of 10%, the rejection region is

$$F < F_{1 - \alpha/2,\, n_1 - 1,\, n_2 - 1} = 1/F_{\alpha/2,\, n_2 - 1,\, n_1 - 1}$$
$$= 1/F_{.05,\, 3,\, 3} = 1/9.28 = .1078$$
$$\text{or } F > F_{\alpha/2,\, n_1 - 1,\, n_2 - 1} = F_{.05,\, 3,\, 3} = 9.28$$

In order for this test to be valid, the populations from which we're sampling must be normally distributed.

EXPERIMENT 6.15

Objective: To illustrate that, when the required condition of the F-test of σ_1^2/σ_2^2 is satisfied, the probability of a Type I error is equal to the significance level.

Experiment: Minitab will generate 1000 pairs of samples from a normal population with a mean of 50 and a standard deviation of 10. For each pair of samples, we will calculate the test statistic and count the number of Type I errors.

Minitab commands:

```
ERASE C1–C49

RANDOM 1000 C1–C8;

NORMAL 50 10.

RSTDEV C1–C4 C9

RSTDEV C5–C8 C10

LET C11 = (C9/C10)**2

CODE (0:.1078)–1(.1078:9.28)0(9.28:9999)1 C11 C12

TABLE C12
```

Columns 9 and 10 contain the sample standard deviations, column 11 stores the values of the test statistic, and column 12 stores the codes, where –1 and 1 indicate that the null hypothesis is rejected and 0 indicates that we do not reject the null hypothesis. Because the null hypothesis is true (the population variances are both 100) the number of –1s and 1s is the number of Type I errors committed.

Instructions for users of full PC versions of Minitab: Type

EXECUTE 'A:EXP6–15'

Instructions for users of student versions of Minitab: Type

EXECUTE 'A:EXP6–15S'

The number of samples to be generated is reduced to 200. We recommend that you repeat this experiment five times and combine the results.

EXPERIMENT 6.16

Objective: To illustrate that, when the populations are nonnormal, the percentage of Type I errors committed in the F-test of σ_1^2/σ_2^2 is not equal to the significance level.

Experiment: We repeat Experiment 6.15, generating the samples from exponential populations whose means (and standard deviations) are equal to 10.

Minitab commands:

ERASE C1–C49

RANDOM 1000 C1–C8;

EXPONENTIAL 10.

RSTDEV C1–C4 C9

RSTDEV C5–C8 C10

LET C11=(C9/C10)**2

CODE (0:.1078)–1(.1078:9.28)0(9.28:9999)1 C11 C12

TABLE C12

Instructions for users of full PC versions of Minitab: Type

EXECUTE 'A:EXP6–16'

Instructions for users of student versions of Minitab: Type

EXECUTE 'A:EXP6–16S'

The number of samples to be generated is reduced to 200. We recommend that you repeat this experiment five times and combine the results.

EXPERIMENT 6.17

Objective: Together with Experiment 6.18, to illustrate the effect of nonnormality on the probability of a Type II error in the F-test of σ_1^2/σ_2^2. In this experiment we estimate the probability of a Type II error when the populations are normal.

Experiment: We repeat Experiment 6.15, except that the population standard deviations are different making the null hypothesis false.

Minitab commands:

```
ERASE C1-C49

RANDOM 1000 C1-C4;

NORMAL 50 10.

RANDOM 1000 C5-C8;

NORMAL 50 20.

RSTDEV C1-C4 C9

RSTDEV C5-C8 C10

LET C11=(C9/C10)**2

CODE (0:.1078)-1(.1078:9.28)0(9.28:9999)1 C11 C12

TABLE C12
```

Columns 1 to 4 contain the observations representing sample 1 (drawn from a normal population whose standard deviation is 10). Columns 5 to 8 contain sample 2 (drawn from a normal population whose standard deviation is 20). The number of 0s in the printout is the number of Type II errors committed.

Instructions for users of full PC versions of Minitab: Type

```
EXECUTE 'A:EXP6-17'
```

Instructions for users of student versions of Minitab: Type

EXECUTE 'A:EXP6–17S'

The number of samples to be generated is reduced to 200. We recommend that you repeat this experiment five times and combine the results.

EXPERIMENT 6.18

Objective: Together with Experiment 6.17, to illustrate the effect of nonnormality on the probability of Type II errors in the F-test of σ_1^2/σ_2^2.

Experiment: We repeat Experiment 6.17 with the data taken from two exponential populations with different standard deviations. Sample 1 is drawn from an exponential distribution whose standard deviation is 10, and sample 2 is drawn from an exponential distribution whose standard deviation is 20.

Minitab commands:

ERASE C1–C49

RANDOM 1000 C1–C4;

EXPONENTIAL 10.

RANDOM C5–C8;

EXPONENTIAL 20.

RSTDEV C1–C4 C9

RSTDEV C5–C8 C10

LET C11 = (C9/C10)**2

CODE (0:.1078)–1(.1078:9.28)0(9.28:9999)1 C11 C12

TABLE C12

Instructions for users of full PC versions of Minitab: Type

EXECUTE 'A:EXP6–18'

Instructions for users of student versions of Minitab: Type

EXECUTE 'A:EXP6–18S'

The number of samples to be generated is reduced to 200. We recommend that you repeat this experiment five times and combine the results.

REPORT FOR EXPERIMENT 6.1

Anticipated number of values of $t < -2.353$:

Anticipated number of values of $t > 2.353$:

Anticipated number of values of t between -2.353 and 2.353:

Anticipated percentage of Type I errors:

Observed number of values of $t < -2.353$:

Observed number of values of $t > 2.353$:

Observed number of values of t between -2.353 and 2.353:

Observed percentage of Type I errors:

(continued)

What do the results of this experiment tell you about the percentage of Type I errors in the t-test of μ when the normality requirement is satisfied?

REPORT FOR EXPERIMENT 6.2

Observed number of values of $t < -2.353$:

Observed number of values of $t > 2.353$:

Observed number of values of t between -2.353 and 2.353:

Observed percentage of Type I errors:

Explain why the observed percentage of Type I errors was greater than 10%.

What do the results of this experiment tell you about the normality requirement for the t-test of μ when the population is moderately nonnormal and the sample is small?

REPORT FOR EXPERIMENT 6.3

Observed number of values of $t < -1.860$:

Observed number of values of $t > 1.860$:

Observed number of values of t between -1.860 and 1.860:

Observed percentage of Type I errors:

What do the results of this experiment and Experiment 6.2 tell you about the effect of the sample size on the importance of the normality requirement of the t-test of μ when the population is moderately nonnormal?

REPORT FOR EXPERIMENT 6.4

Observed number of values of $t < -1.711$:

Observed number of values of $t > 1.711$:

Observed number of values of t between -1.711 and 1.711:

Observed percentage of Type I errors:

What conclusions can you draw from Experiments 6.2—6.4 regarding the effect of the sample size on the importance of the normality requirement of the t-test of μ when the population is moderately nonnormal?

(*continued*)

Can you provide a rough guide to specify the minimum sample size required to ensure the validity of the t-test of μ when the population is moderately nonnormal?

REPORT FOR EXPERIMENT 6.5

Observed number of values of $t < -2.353$:

Observed number of values of $t > 2.353$:

Observed number of values of t between -2.353 and 2.353:

Observed percentage of Type I errors:

Explain why the observed percentage of Type I errors was greater than 10%.

(*continued*)

Why do you suppose that there were more Type I errors occurring in the left tail of the sampling distribution ($t < -2.353$) than in the right tail of the sampling distribution ($t > 2.353$) of the test statistic? What does this tell you about the shape of the sampling distribution of the test statistic?

REPORT FOR EXPERIMENT 6.6

Observed number of values of $t < -1.860$:

Observed number of values of $t > 1.860$:

Observed number of values of t between -1.860 and 1.860:

Observed percentage of Type I errors:

What do the results of this experiment and Experiment 6.5 tell you about the effect of the sample size on the importance of the normality requirement of the t-test of μ when the population is extremely nonnormal?

REPORT FOR EXPERIMENT 6.7

Observed number of values of $t < -1.711$:

Observed number of values of $t > 1.711$:

Observed number of values of t between -1.711 and 1.711:

Observed percentage of Type I errors:

What conclusions can you draw from Experiments 6.5—6.7 regarding the effect of the sample size on the importance of the normality requirement of the t-test of μ when the population is extremely nonnormal?

Can you provide a rough guide to specify the minimum sample size necessary to ensure the validity of the t-test of μ when the population is extremely nonnormal?

REPORT FOR EXPERIMENT 6.8

Observed percentage of Type II errors:

REPORT FOR EXPERIMENT 6.9

Observed percentage of Type II errors:

What do the results of Experiments 6.8 and 6.9 tell you about the percentage of Type II errors in the t-test of $(\mu_1 - \mu_2)$ when the populations are normal and moderately nonnormal?

Can you generalize the results of Experiments 6.8 and 6.9 to situations in which the population is extremely nonnormal?

REPORT FOR EXPERIMENT 6.10

Anticipated percentage of Type I errors:

Observed percentage of Type I errors:

What do the results of this experiment tell you about the percentage of Type I errors in the t-test of $(\mu_1 - \mu_2)$ when the required conditions are satisfied?

REPORT FOR EXPERIMENT 6.11

Anticipated percentage of Type I errors:

(*continued*)

Observed percentage of Type I errors:

What do the results of Experiment 6.10 and 6.11 tell you about the importance of the equal-variance requirement of the t-test of $(\mu_1 - \mu_2)$ when the variances are distinctly different?

REPORT FOR EXPERIMENT 6.12

Anticipated percentage of Type I errors:

Observed percentage of Type I errors:

What do the results of Experiment 6.10 and 6.12 tell you about the importance of the normality requirement of the t-test of $(\mu_1 - \mu_2)$?

REPORT FOR EXPERIMENT 6.13

Observed percentage of Type II errors:

REPORT FOR EXPERIMENT 6.14

Observed percentage of Type II errors:

What do the results of Experiments 6.13 and 6.14 tell you about the percentage of Type II errors in the t-test of $(\mu_1 - \mu_2)$ when the populations are normal and nonnormal?

(continued)

Briefly summarize your findings from Experiments 6.10 and 6.12–6.14. In particular, comment on the advisability of using the t-test of $(\mu_1 - \mu_2)$ when the populations are nonnormal.

REPORT FOR EXPERIMENT 6.15

Anticipated percentage of Type I errors:

Observed percentage of Type I errors:

What do the results of this experiment tell you about the percentage of Type I errors in the F-test of σ_1^2/σ_2^2 when the required condition is satisfied?

REPORT FOR EXPERIMENT 6.16

Anticipated percentage of Type I errors:

(continued)

Observed percentage of Type I errors:

What do the results of Experiment 6.15 and 6.16 tell you about the importance of the normality requirement of the F-test of σ_1^2/σ_2^2?

REPORT FOR EXPERIMENT 6.17

Observed percentage of Type II errors:

REPORT FOR EXPERIMENT 6.18

Observed percentage of Type II errors:

What do the results of Experiments 6.17 and 6.18 tell you about the percentage of Type II errors in the F-test of σ_1^2/σ_2^2 when the populations are normal and nonnormal?

(continued)

Briefly summarize your findings from Experiments 6.15—6.18. In particular, comment on the advisability of using the F-test of σ_1^2/σ_2^2 when the populations are nonnormal.

CHAPTER 7

Statistical Process Control

7.1 INTRODUCTION

Statistical process control (SPC) is an application of hypothesis testing with two major differences. First, in the "traditional" test of hypothesis, we test for an unknown but fixed value of a parameter. In SPC we test to determine whether the distribution of a dynamic process has changed. Second, in traditional tests we take one sample, calculate the test statistic, and base our decision on its value. In SPC we determine whether the process is out of control by observing the values of a series of statistics. The statistics are plotted on a control chart and several tests can be applied.

To demonstrate the concepts of SPC, we will describe experiments involving the construction of $\bar{x}$ charts. Minitab applies eight different tests to determine whether the mean of the production process has changed. These tests are:

1. One point beyond zone A
2. Nine points in a row in zone C or beyond (on one side of the centerline)
3. Six points in a row, all increasing or all decreasing
4. Fourteen points in a row, alternating up and down
5. Two out of three points in a row in zone A or beyond (on one side of the centerline)
6. Four out of five points in a row in zone B or beyond (on one side of the centerline)
7. Fifteen points in a row in zones C (above and below the centerline)
8. Eight points in a row beyond zones C (above and below the centerline)

The purpose of the four experiments presented in this chapter is to illustrate Type I and Type II errors in SPC.

7.2 TYPE I AND TYPE II ERRORS

A Type I error occurs when one of the tests indicates that the process is out of control when in fact it is not. A Type II error occurs when the process is out of control and none of the tests indicate this condition. Because SPC techniques are based on a series of samples, errors can be made for each sample.

EXPERIMENT 7.1

Objective: To illustrate that statistical process control techniques will commit Type I errors in the same way as the "traditional" tests of hypotheses do.

Experiment: We will generate 400 samples of size 4 from one normal distribution whose mean and standard deviation do not change ($\mu = 50$ and $\sigma = 10$). Minitab will be instructed to construct an $\bar{x}$ chart. Because the distribution of the process does not change, any tests that indicate that the process is out of control will commit a Type I error.

Minitab commands:

> *ERASE C1–C49*
>
> *RANDOM 1600 C1;*
>
> *NORMAL 50 10.*
>
> *XBAR C1 4;*
>
> *TEST 1:8.*

We generate 1600 observations from a normal distribution whose mean and standard deviation are 50 and 10, respectively, and store the data in column 1. The command *XBAR C1 4* instructs Minitab to treat the 1600 observations as 400 samples of size 4. The subcommand *TEST 1:8* tells Minitab to conduct all eight tests. Note that the $\bar{x}$ chart drawn by Minitab is limited to 50 samples. Thus, when drawing an $\bar{x}$ chart with 400 samples, Minitab outputs the chart in 8 segments of 50 samples. The results of the tests are summarized at the end.

Instructions for all users of Minitab: Type

> *EXECUTE 'A:EXP7–1'*

Suppose that we know that, when a production process is under control, the mean and standard deviation are 50 and 10, respectively. (In practice these values would be estimated from previous control charts.) The next three experiments illustrate what happens when the process mean is not equal to 50. Because the mean has changed, we expect that the $\bar{x}$ chart would indicate that the process is out of control. However, not all sample means will indicate that the process is out of control, thus committing a Type II error. The following experiments will illustrate Type II errors for different process distribution means.

EXPERIMENT 7.2

Objective: To illustrate that, when the process is out of control, a large number of samples will erroneously conclude that the process is under control.

Experiment: We will generate 400 samples of size 4 from a normal population whose mean is 51 (and whose standard deviation is 10). Minitab will then create the $\bar{x}$ chart assuming a mean of 50 and a standard deviation of 10. All eight tests will be conducted.

Minitab commands:

> ERASE C1–C49
>
> RANDOM 1600 C1;
>
> NORMAL 51 10.
>
> XBAR C1 4;
>
> MU = 50;
>
> SIGMA = 10;
>
> TEST 1:8.

The subcommands *MU = 50* and *SIGMA = 10* tell Minitab to construct the $\bar{x}$ chart assuming these values for the mean and standard deviation. Because the mean of the data is actually 51, the mean has changed and the process is out of control. The objective is to count the number of samples (out of the 400 samples generated) that do not indicate that the process is out of control. Also, count the number of samples tested until the correct conclusion is drawn.

Instructions for all users of Minitab: Type

> EXECUTE 'A:EXP7–2'

EXPERIMENT 7.3

Objective: In concert with Experiments 7.2 and 7.4, to show the effect of different values of the process mean on the occurrence of Type II errors in $\bar{x}$ charts.

Experiment: We repeat Experiment 7.2, changing the mean to 52.

Minitab commands:

> *ERASE C1–C49*
>
> *RANDOM 1600 C1;*
>
> *NORMAL 52 10.*
>
> *XBAR C1 4;*
>
> *MU = 50;*
>
> *SIGMA = 10;*
>
> *TEST 1:8.*

Instructions for all users of Minitab: Type

> *EXECUTE 'A:EXP7–3'*

EXPERIMENT 7.4

Objective: In concert with Experiments 7.2 and 7.3, to show the effect of different values of the process mean on the occurrence of Type II errors in $\bar{x}$ charts.

Experiment: We repeat Experiment 7.2, changing the mean to 55.

Minitab commands:

> *ERASE C1–C49*
>
> *RANDOM 1600 C1;*
>
> *NORMAL 55 10.*
>
> *XBAR C1 4;*
>
> *MU = 50;*
>
> *SIGMA = 10;*
>
> *TEST 1:8.*

Instructions for all users of Minitab: Type

EXECUTE 'A:EXP7–4'

REPORT FOR EXPERIMENT 7.1

Number of samples that resulted in the (correct) decision that the process is under control:

Number of samples that resulted in a Type I error (concluding that the process is out of control):

How many Type I errors did you anticipate seeing using test 1?

How many Type I errors did you observe using test 1?

Is it possible to calculate the probability of Type I errors using the other tests? If so, calculate these probabilities and compare them to the number of Type I errors committed by these tests.

REPORT FOR EXPERIMENT 7.2

Number of samples that resulted in a Type II error (not concluding that the process is out of control):

Number of samples (counting from the point when the process went out of control—sample 1) until the $\bar{x}$ chart determined that the process is now out of control:

Discuss why there is such a large number of Type II errors in this experiment, and why the control chart requires such a large number of samples before recognizing that the process is out of control.

REPORT FOR EXPERIMENT 7.3

Number of samples that resulted in a Type II error (not concluding that the process is out of control):

Number of samples (counting from the point when the process went out of control—before sample 1 was taken) until the $\bar{x}$ chart determined that the process is out of control:

REPORT FOR EXPERIMENT 7.4

Number of samples that resulted in a Type II error (not concluding that the process is out of control):

Number of samples until the $\bar{x}$ chart determined that the process is out of control:

(continued)

Summarize the results of the four experiments described in this chapter. Also, briefly discuss the effects of changing the process mean on the occurrence of Type II errors on $\bar{x}$ charts.

CHAPTER 8

Simple Linear Regression

8.1 INTRODUCTION

The simple linear regression model is expressed as

$$y = \beta_0 + \beta_1 x + \epsilon$$

where y is the dependent variable, x is the independent variable, β_0 is the y-intercept, β_1 is the slope, and ϵ is the error variable. The error variable is assumed to be normally distributed with a mean of zero and a constant (and usually unknown) variance, which we denote σ_ϵ^2.

We estimate the parameters β_0 and β_1 by drawing a sample of size n and finding the coefficients that produce the "best" straight line. The best line is defined as the one that minimizes the sum of squared errors. The errors are defined as the differences between the points and the line. The sum of squared errors is denoted SSE. The estimators are labeled $\hat{\beta}_0$ and $\hat{\beta}_1$. There are a variety of statistical procedures to assess how well the model fits. The three most commonly used are listed below:

1. The estimate of the error variable variance: The statistic s_ϵ^2 (defined as SSE$/(n - 2)$ is an unbiased estimator of the variance of the error variable. Its square root, s_ϵ is called the standard error of estimate. (The standard error of estimate is not an unbiased estimator of σ_ϵ.)

2. t-test of β_1: We test to determine whether there is enough evidence to infer that β_1 is different from zero. (A slope of zero indicates that no linear relationship exists.) The hypotheses are

    $$H_0: \beta_1 = 0$$

    $$H_A: \beta_1 \neq 0$$

    The test statistic is

    $$t = \frac{\hat{\beta}_1 - \beta_1}{s_{\hat{\beta}_1}}$$

which is Student t distributed with $n - 2$ degrees of freedom. The rejection region is

$$|t| > t_{\alpha/2,\, n - 2}$$

3. Coefficient of determination: The statistic r^2, which lies between 0 and 1, measures the proportion of the variation in y that is explained by the variation in x.

In this chapter we present several experiments designed to illustrate important concepts associated with regression analysis. In Section 8.2 we describe experiments that illustrate the effect of increasing the variance of the error variable when there is a linear relationship. Section 8.3 investigates confidence interval estimates of the expected value of y. Section 8.4 illustrates what happens when the sample size changes, and Section 8.5 describes experiments that illustrate how the spread of x affects the results.

8.2 EFFECT OF THE ERROR VARIABLE VARIANCE

In this section we assume that the model representing the relationship between two variables is

$$y = 10 + 5x + \epsilon$$

That is, the true value of β_0 is 10 and the true value of β_1 is 5. If the error variable is small, which means that its variance is small, the regression line fits quite well. If, however, the errors are large, which means that its variance is large, the regression line's fit is quite poor. The three experiments that follow address the question of what happens to the regression statistics as the model's fit worsens.

EXPERIMENT 8.1

Objective: Together with Experiments 8.2 and 8.3, to show how the regression statistics are affected by increasing the value of σ_ϵ^2.

Experiment: We generate 500 samples of size 5. Each sample consists of the values $x = 1, 2, 3, 4$, and 5 and their associated values of y. That is, the samples are

x	$y = 10 + 5x + \epsilon$
1	$15 + \epsilon$
2	$20 + \epsilon$
3	$25 + \epsilon$

| 4 | $30 + \epsilon$ |
| 5 | $35 + \epsilon$ |

The values of ϵ will be generated from a normal distribution whose mean is 0 and whose variance is 1 (and $\sigma_\epsilon = 1$). For each sample, we will compute the relevant statistics $\hat{\beta}_1$, $\hat{\beta}_0$, s_ϵ^2, t, and r^2. The descriptive statistics and the histograms for these five statistics will be outputted.

Minitab commands:

ERASE C1–C49

RANDOM 500 C1–C5;

NORMAL 0 1.

LET C6=C1+15

LET C7=C2+20

LET C8=C3+25

LET C9=C4+30

LET C10=C5+35

LET C11=C6*1

LET C12=C7*2

LET C13=C8*3

LET C14=C9*4

LET C15=C10*5

ERASE C1–C5

RSUM C6–C10 C1

RSSQ C6–C10 C2

RSUM C11–C15 C3

ERASE C6–C15

```
LET C4=C3-3*C1

LET C5=C2-C1*C1/5

LET C6=C4/10

LET C7=C1/5-C6*3

LET C8=C5-C4*C4/10

LET C9=C8/3

LET C10=(C6*3.1623)/C9

LET C11=C4*C4/(10*C5)

NAME C6='BETA-1'

NAME C7='BETA-0'

NAME C9='VAR-EST'

NAME C10='t'

NAME C11='R-SQ'

DESCRIBE C6

HISTOGRAM C6

DESCRIBE C7

HISTOGRAM C7

DESCRIBE C9

HISTOGRAM C9

DESCRIBE C10

HISTOGRAM C10

DESCRIBE C11

HISTOGRAM C11
```

CODE (–9999:–2.353)–1(–2.353:2.353)0(2.353:9999)1 C10 C12

TABLE C12

The normally distributed errors are stored in columns 1 through 5, and the values of y are stored in columns 6 through 10. After erasing (to make more workspace) columns 1 to 5, we store various sums and sums of squares in the following columns:

Column 1 = Σy

Column 2 = Σy^2

Column 3 = Σxy

Column 4 = $SS_{xy} = \Sigma(x - \bar{x})(y - \bar{y}) = \Sigma xy - (\Sigma x)(\Sigma y)/n$

Column 5 = $SS_y = \Sigma(y - \bar{y})^2 = \Sigma y^2 - (\Sigma y)^2/n$

Note that in these computations and in the ones that follow, we use

$\Sigma x = 15$

$\Sigma x^2 = 55$

$SS_x = \Sigma(x - \bar{x})^2 = \Sigma x^2 - (\Sigma x)^2/n = 10$

We erase columns 6 to 15, and calculate and store the regression statistics in the following columns:

Column 6 = $\hat{\beta}_1$

Column 7 = $\hat{\beta}_0$

Column 8 = SSE

Column 9 = s_ϵ^2

Column 10 = t

Column 11 = r^2

The output includes the descriptive statistics and histograms for these statistics (except *SSE*). As we've done in other chapters, the values of t are coded to report the number of tests that lead to the rejection (-1 and $+1$) and the nonrejection (0) of the null hypothesis, H_0: $\beta_1 = 0$. The rejection region (with $\alpha = .10$) is

$$|t| > 2.353$$

Since the null hypothesis is false, a t-statistic that lies between -2.353 and 2.353 represents a Type II error.

Instructions for users of full PC versions of Minitab: Type

EXECUTE 'A:EXP8-1'

Instructions for users of student versions of Minitab: (Because of DOS limitations, it was necessary to store all eight student versions of the program in this chapter in a separate subdirectory. Note the different instruction.) Type:

EXECUTE 'A:C8\EXP8-1S'

The number of samples to be generated is reduced to 100.

EXPERIMENT 8.2

Objective: Together with Experiments 8.1 and 8.3, to show how the regression statistics are affected by increasing the value of σ_ϵ^2.

Experiment: We repeat Experiment 8.1 with $\sigma_\epsilon^2 = 25$ $(\sigma_\epsilon = 5)$.

Minitab commands:

ERASE C1-C49

RANDOM 500 C1-C5;

NORMAL 0 5.

LET C6=C1+15

LET C7=C2+20

LET C8=C3+25

LET C9=C4+30

```
LET C10=C5+35

LET C11=C6*1

LET C12=C7*2

LET C13=C8*3

LET C14=C9*4

LET C15=C10*5

ERASE C1–C5

RSUM C6–C10 C1

RSSQ C6–C10 C2

RSUM C11–C15 C3

ERASE C6–C15

LET C4=C3–3*C1

LET C5=C2–C1*C1/5

LET C6=C4/10

LET C7=C1/5–C6*3

LET C8=C5–C4*C4/10

LET C9=C8/3

LET C10=(C6*3.1623)/C9

LET C11=C4*C4/(10*C5)

NAME C6='BETA–1'

NAME C7='BETA–0'

NAME C9='VAR–EST'

NAME C10='t'
```

NAME C11 = 'R-SQ'

DESCRIBE C6

HISTOGRAM C6

DESCRIBE C7

HISTOGRAM C7

DESCRIBE C9

HISTOGRAM C9

DESCRIBE C10

HISTOGRAM C10

DESCRIBE C11

HISTOGRAM C11

CODE (-999:-2.353)-1(-2.353:2.353)0(2.353:999)1 C10 C12

TABLE C12

Instructions for users of full PC versions of Minitab: Type

EXECUTE 'A:EXP8-2'

Instructions for users of student versions of Minitab: Type

EXECUTE 'A:C8\EXP8-2S'

The number of samples to be generated is reduced to 100.

EXPERIMENT 8.3

Objective: Together with Experiments 8.1 and 8.2, to show how the regression statistics are affected by increasing the value of σ_ϵ^2.

Experiment: We repeat Experiment 8.1 with $\sigma_\epsilon^2 = 625$ $(\sigma_\epsilon = 25)$.

Minitab commands:

```
ERASE C1–C49

RANDOM 500 C1–C5;

NORMAL 0 25.

LET C6=C1+15

LET C7=C2+20

LET C8=C3+25

LET C9=C4+30

LET C10=C5+35

LET C11=C6*1

LET C12=C7*2

LET C13=C8*3

LET C14=C9*4

LET C15=C10*5

ERASE C1–C5

RSUM C6–C10 C1

RSSQ C6–C10 C2

RSUM C11–C15 C3

ERASE C6–C15

LET C4=C3–3*C1

LET C5=C2–C1*C1/5

LET C6=C4/10

LET C7=C1/5–C6*3
```

LET C8=C5-C4*C4/10

LET C9=C8/3

LET C10=(C6*3.1623)/C9

LET C11=C4*C4/(10*C5)

NAME C6='BETA-1'

NAME C7='BETA-0'

NAME C9='VAR-EST'

NAME C10='t'

NAME C11='R-SQ'

DESCRIBE C6

HISTOGRAM C6

DESCRIBE C7

HISTOGRAM C7

DESCRIBE C9

HISTOGRAM C9

DESCRIBE C10

HISTOGRAM C10

DESCRIBE C11

HISTOGRAM C11

CODE (-999:-2.353)-1(-2.353:2.353)0(2.353:999)1 C10 C12

TABLE C12

Instructions for users of full PC versions of Minitab: Type

EXECUTE 'A:EXP8-3'

Instructions for users of student versions of Minitab: Type

EXECUTE 'A:C8\EXP8–3S'

The number of samples to be generated is reduced to 100.

8.3 ESTIMATING THE EXPECTED VALUE OF *y*

One of the purposes of regression analysis is to produce an estimate of the expected value of the dependent variable for a given value of the independent variable. The confidence interval estimator is

$$\hat{y} \pm t_{\alpha/2} s_\epsilon \sqrt{\frac{1}{n} + \frac{(x_g - \bar{x})^2}{SS_x}}$$

where x_g is the given value of x. The effect of the term

$$\frac{(x_g - \bar{x})^2}{SS_x}$$

is that when the difference between x_g and $\bar{x}$ is large, the interval estimate will be wider than when the difference between x_g and $\bar{x}$ is small. The purpose of the next experiment is to illustrate this effect.

EXPERIMENT 8.4

Objective: To illustrate the effect of different values of x_g on the width of the confidence interval estimator of the expected value of y.

Experiment: We repeat Experiment 8.2, but instead of printing the statistics used to assess the model, we output the point estimate of the expected value of y when x_g takes on the values 3, 5, and 7.

Minitab commands:

ERASE C1–C49

RANDOM 500 C1–C5;

NORMAL 0 5.

LET C6=C1+15

```
LET C7=C2+20

LET C8=C3+25

LET C9=C4+30

LET C10=C5+35

LET C11=C6*1

LET C12=C7*2

LET C13=C8*3

LET C14=C9*4

LET C15=C10*5

ERASE C1–C5

RSUM C6–C10 C1

RSSQ C6–C10 C2

RSUM C11–C15 C3

ERASE C6–C15

LET C4=C3–3*C1

LET C5=C2–C1*C1/5

LET C6=C4/10

LET C7=C1/5–C6*3

LET C8=C7+3*C6

DESCRIBE C8

HISTOGRAM C8

LET C9=C7+5*C6

DESCRIBE C9
```

HISTOGRAM C9

LET C10=C7+7*C6

DESCRIBE 10

HISTOGRAM C10

For each sample, we calculate $\hat{\beta}_1$ and $\hat{\beta}_0$ (columns 6 and 7, respectively). From these estimates we compute

$$\hat{y} = \hat{\beta}_0 + \hat{\beta}_1 x_g$$

The values of $\hat{y}$ when $x_g = 3$ are stored in column 8, values of $\hat{y}$ when $x_g = 5$ are stored in column 9, and values of $\hat{y}$ when $x_g = 7$ are stored in column 10. Minitab prints the descriptive statistics and histograms for each set of values.

Instructions for users of full PC versions of Minitab: Type

EXECUTE 'A:EXP8–4'

Instructions for users of student versions of Minitab: Type

EXECUTE 'A:C8\EXP8–4S'

The number of samples to be generated is reduced to 100.

8.4   EFFECT OF THE SAMPLE SIZE

In earlier chapters we examined the effect of the sample size on several inferential techniques. In this section we describe two experiments that illustrate what happens to the regression analysis when the sample size is changed.

EXPERIMENT 8.5

Objective: Together with experiments 8.2 and 8.6, to illustrate the effect of the sample size on the regression statistics.

Experiment: Experiment 8.2 is repeated using a sample size of 10. The x-values are 1, 2, . . . , 10.

Minitab commands:

ERASE C1–C49

RANDOM 500 C1–C10;

NORMAL 0 5.

LET C11 = C1 + 15

LET C12 = C2 + 20

LET C13 = C3 + 25

LET C14 = C4 + 30

LET C15 = C5 + 35

LET C16 = C6 + 40

LET C17 = C7 + 45

LET C18 = C8 + 50

LET C19 = C9 + 55

LET C20 = C10 + 60

LET C21 = C11*1

LET C22 = C12*2

LET C23 = C13*3

LET C24 = C14*4

LET C25 = C15*5

LET C26 = C16*6

LET C27 = C17*7

LET C28 = C18*8

LET C29 = C19*9

```
LET C30=C20*10

ERASE C1–C10

RSUM C11–C20 C1

RSSQ C11–C20 C2

RSUM C21–C30 C3

ERASE C11–C30

LET C4=C3–5.5*C1

LET C5=C2–C1*C1/10

LET C6=C4/82.5

LET C7=C1/10–C6*5.5

LET C8=C5–C4*C4/82.5

LET C9=C8/8

LET C10=(C6*9.083)/C9

LET C11=C4*C4/(82.5*C5)

NAME C6='BETA–1'

NAME C7='BETA–0'

NAME C9='VAR–EST'

NAME C10='t'

NAME C11='R–SQ'

DESCRIBE C6

HISTOGRAM C6

DESCRIBE C7

HISTOGRAM C7
```

DESCRIBE C9

HISTOGRAM C9

DESCRIBE C10

HISTOGRAM C10

DESCRIBE C11

HISTOGRAM C11

CODE (–999:–1.860)–1(–1.860:1.860)0(1.860:999)1 C10 C12

TABLE C12

The 10 values of ϵ are generated and stored in columns 1 through 10. The values of y calculated from the equation

$$y = 10 + 5x + \epsilon$$

are stored in columns 11 through 20. As in Experiment 8.2, we compute the sums and sums of squares (using $\Sigma x = 55$, $\Sigma x^2 = 385$, and $SS_x = 82.5$) to determine the relevant statistics. The rejection region of the test of β_1 (with $\alpha = .10$) is

$$|t| > 1.860$$

Instructions for users of full PC versions of Minitab: Type

EXECUTE 'A:EXP8-5'

Instructions for users of student versions of Minitab: Type

EXECUTE 'A:C8\EXP8-5S'

The number of samples to be generated is reduced to 100.

EXPERIMENT 8.6

Objective: Together with Experiments 8.2 and 8.5, to illustrate the effect of the sample size.

Experiment: Experiment 8.2 is repeated using a sample size of 15. (The number of samples is reduced to 250 because of workspace limitations.) The x-values are 1, 2, ..., 15 (and $\Sigma x = 120$, $\Sigma x^2 = 1240$, and $SS_x = 280$). The rejection region of the test of the slope (with $\alpha = .10$) is

$$|t| > 1.771$$

Minitab commands:

```
ERASE C1–C49

RANDOM 250 C1–C15;

NORMAL 0 5.

LET C16=C1+15

LET C17=C2+20

LET C18=C3+25

LET C19=C4+30

LET C20=C5+35

LET C21=C6+40

LET C22=C7+45

LET C23=C8+50

LET C24=C9+55

LET C25=C10+60

LET C26=C11+65

LET C27=C12+70

LET C28=C13+75

LET C29=C14+80

LET C30=C15+85

LET C31=C16*1
```

LET C32=C17*2

LET C33=C18*3

LET C34=C19*4

LET C35=C20*5

LET C36=C21*6

LET C37=C22*7

LET C38=C23*8

LET C39=C24*9

LET C40=C25*10

LET C41=C26*11

LET C42=C27*12

LET C43=C28*13

LET C44=C29*14

LET C45=C30*15

ERASE C1–C15

RSUM C16–C30 C1

RSSQ C16–C30 C2

RSUM C31–C45 C3

ERASE C16–C45

LET C4=C3–8*C1

LET C5=C2–C1*C1/15

LET C6=C4/280

LET C7=C1/15–C6*8

LET C8=C5–C4*C4/280

LET C9=C8/13

LET C10=(C6*16.7332)/C9

LET C11=C4*C4/(280*C5)

NAME C6='BETA–1'

NAME C7='BETA–0'

NAME C9='VAR–ST'

NAME C10='t'

NAME C11='R-SQ'

DESCRIBE C6

HISTOGRAM C6

DESCRIBE C7

HISTOGRAM C7

DESCRIBE C9

HISTOGRAM C9

DESCRIBE C10

HISTOGRAM C10

DESCRIBE C11

HISTOGRAM C11

CODE (–999:–1.771)–1(–1.771:1.771)0(1.771:999)1 C10 C12

TABLE C12

Instructions for users of full PC versions of Minitab: Type

EXECUTE 'A:EXP8–6'

Instructions for users of student versions of Minitab: Type

> *EXECUTE 'A:C8\EXP8–6S'*

The number of samples to be generated is reduced to 50.

8.5 EFFECT OF INCREASING THE SPREAD OF x

In many realistic applications of regression analysis, the statistician controls the values of the independent variable. For example, in an analysis of the relationship between income and education, a statistician may choose a random sample of people and ask each to report his or her annual income and the number of years of formal education. Alternatively, the statistician may select individuals with specific amounts of education and then ask each of them how much they earn. The next two experiments are designed to illustrate an important concept when setting up experiments to estimate regression models.

EXPERIMENT 8.7

Objective: Together with Experiment 8.8, to show how the regression statistics are affected by increasing the spread of x.

Experiment: We repeat Experiment 8.2 with $x = 8, 9, 10, 11,$ and 12 (with $\Sigma x = 50$, $\Sigma x^2 = 510$, and $SS_x = 10$).

Minitab commands:

> *ERASE C1–C49*
>
> *RANDOM 500 C1–C5;*
>
> *NORMAL 0 5.*
>
> *LET C6=C1+50*
>
> *LET C7=C2+55*
>
> *LET C8=C3+60*
>
> *LET C9=C4+65*
>
> *LET C10=C5+70*
>
> *LET C11=C6*8*

LET C12=C7*9

LET C13=C8*10

LET C14=C9*11

LET C15=C10*12

ERASE C1-C5

RSUM C6-C10 C1

RSSQ C6-C10 C2

RSUM C11-C15 C3

ERASE C6-C15

LET C4=C3-10*C1

LET C5=C2-C1*C1/5

LET C6=C4/10

LET C7=C1/5-C6*10

LET C8=C5-C4*C4/10

LET C9=C8/3

LET C10=(C6*3.1623)/C9

LET C11=C4*C4/(10*C5)

NAME C6='BETA-1'

NAME C7='BETA-0'

NAME C9='VAR-EST'

NAME C10='t'

NAME C11='R-SQ'

DESCRIBE C6

HISTOGRAM C6

DESCRIBE C7

HISTOGRAM C7

DESCRIBE C9

HISTOGRAM C9

DESCRIBE C10

HISTOGRAM C10

DESCRIBE C11

HISTOGRAM C11

CODE (–999:–2.353)–1(–2.353:2.353)0(2.353:999)1 C10 C12

TABLE C12

Instructions for users of full PC versions of Minitab: Type

EXECUTE 'A:EXP8–7'

Instructions for users of student versions of Minitab: Type

EXECUTE 'A:C8\EXP8–7S'

The number of samples to be generated is reduced to 100.

EXPERIMENT 8.8

Objective: Together with Experiment 8.7, to show how the regression statistics are affected by increasing the spread of x.

Experiment: We repeat Experiment 8.2 with $x = 0, 5, 10, 15,$ and 20 (and $\Sigma x = 50$, $\Sigma x^2 = 750$, and $SS_x = 250$).

Minitab commands:

ERASE C1–C49

RANDOM 500 C1–C5;

NORMAL 0 5.

LET C6=C1+10

LET C7=C2+35

LET C8=C3+60

LET C9=C4+85

LET C10=C5+110

LET C11=C6*0

LET C12=C7*5

LET C13=C8*10

LET C14=C9*15

LET C15=C10*20

ERASE C1–C5

RSUM C6–C10 C1

RSSQ C6–C10 C2

RSUM C11–C15 C3

ERASE C6–C15

LET C4=C3–10*C1

LET C5=C2–C1*C1/5

LET C6=C4/250

LET C7=C1/5–C6*10

LET C8=C5–C4*C4/250

LET C9=C8/3

LET C10=(C6*15.8114)/C9

LET C11 = C4*C4/(250*C5)

NAME C6 = 'BETA-1'

NAME C7 = 'BETA-0'

NAME C9 = 'VAR-EST'

NAME C10 = 't'

NAME C11 = 'R-SQ'

DESCRIBE C6

HISTOGRAM C6

DESCRIBE C7

HISTOGRAM C7

DESCRIBE C9

HISTOGRAM C9

DESCRIBE C10

HISTOGRAM C10

DESCRIBE C11

HISTOGRAM C11

CODE (–999:–2.353)–1(–2.353:2.353)0(2.353:999)1 C10 C12

TABLE C12

Instructions for users of full PC versions of Minitab: Type

EXECUTE 'A:EXP8–8'

Instructions for users of student versions of Minitab: Type

EXECUTE 'A:C8\EXP8–8S'

The number of samples to be generated is reduced to 100.

REPORT FOR EXPERIMENT 8.1

What values for the means of $\hat{\beta}_1$, $\hat{\beta}_0$, and s_ϵ^2 did you anticipate seeing?

What values for the means of $\hat{\beta}_1$, $\hat{\beta}_0$, and s_ϵ^2 did you observe?

What values for the standard deviations of $\hat{\beta}_1$, $\hat{\beta}_0$, s_ϵ^2, t, and r^2 did you observe?

Percentage of Type II errors:

REPORT FOR EXPERIMENT 8.2

What values for the means of $\hat{\beta}_1$, $\hat{\beta}_0$, and s_ϵ^2 did you anticipate seeing?

What values for the means of $\hat{\beta}_1$, $\hat{\beta}_0$, and s_ϵ^2 did you observe?

What values for the standard deviations of $\hat{\beta}_1$, $\hat{\beta}_0$, s_ϵ^2, t, and r^2 did you observe?

Percentage of Type II errors:

REPORT FOR EXPERIMENT 8.3

What values for the means of $\hat{\beta}_1$, $\hat{\beta}_0$, and s_ϵ^2 did you anticipate seeing?

What values for the means of $\hat{\beta}_1$, $\hat{\beta}_0$, and s_ϵ^2 did you observe?

What values for the standard deviations of $\hat{\beta}_1$, $\hat{\beta}_0$, s_ϵ^2, t, and r^2 did you observe?

Percentage of Type II errors:

(continued)

Briefly describe what happens to the distributions of $\hat{\beta}_1$, $\hat{\beta}_0$, s_ϵ^2, t, and r^2 and the percentage of Type II errors as σ_ϵ^2 increases. What are the implications of your findings?

REPORT FOR EXPERIMENT 8.4

What value for the mean of the expected value of y when $x = 3$ did you anticipate seeing?

What value for the mean of the expected value of y when $x = 3$ did you observe?

What value for the standard deviation of the expected value of y when $x = 3$ did you observe?

What value for the mean of the expected value of y when $x = 5$ did you anticipate seeing?

What value for the mean of the expected value of y when $x = 5$ did you observe?

What value for the standard deviation of the expected value of y when $x = 5$ did you observe?

What value for the mean of the expected value of y when $x = 7$ did you anticipate seeing?

(continued)

What value for the mean of the expected value of y when $x = 7$ did you observe?

What value for the standard deviation of the expected value of y when $x = 7$ did you observe?

Briefly describe what the results of this experiment tell you about the relationship between the value of x_g and the width of the interval estimate. Why does the interval widen as x_g moves farther away from $\bar{x}$?

REPORT FOR EXPERIMENT 8.5

What values for the means of $\hat{\beta}_1$, $\hat{\beta}_0$, and s_ϵ^2 did you anticipate seeing?

What values for the means of $\hat{\beta}_1$, $\hat{\beta}_0$, and s_ϵ^2 did you observe?

What values for the standard deviations of $\hat{\beta}_1$, $\hat{\beta}_0$, s_ϵ^2, t, and r^2 did you observe?

Percentage of Type II errors:

REPORT FOR EXPERIMENT 8.6

What values for the means of $\hat{\beta}_1$, $\hat{\beta}_0$, and s_ϵ^2 did you anticipate seeing?

What values for the means of $\hat{\beta}_1$, $\hat{\beta}_0$, and s_ϵ^2 did you observe?

What values for the standard deviations of $\hat{\beta}_1$, $\hat{\beta}_0$, s_ϵ^2, t, and r^2 did you observe?

Percentage of Type II errors:

(continued)

Summarize the results of Experiments 8.2, 8.5, and 8.6. Discuss the effect that increasing the sample size has on the distributions of $\hat{\beta}_1$, $\hat{\beta}_0$, s_ϵ^2, t, and r^2.

REPORT FOR EXPERIMENT 8.7

What values for the means of $\hat{\beta}_1$, $\hat{\beta}_0$, and s_ϵ^2 did you anticipate seeing?

What values for the means of $\hat{\beta}_1$, $\hat{\beta}_0$, and s_ϵ^2 did you observe?

What values for the standard deviations of $\hat{\beta}_1$, $\hat{\beta}_0$, s_ϵ^2, t, and r^2 did you observe?

Percentage of Type II errors:

(continued)

Discuss the differences between the results of this experiment and those of Experiment 8.2. How do you explain the differences?

REPORT FOR EXPERIMENT 8.8

What values for the means of $\hat{\beta}_1$, $\hat{\beta}_0$, and s_ϵ^2 did you anticipate seeing?

What values for the means of $\hat{\beta}_1$, $\hat{\beta}_0$, and s_ϵ^2 did you observe?

What values for the standard deviations of $\hat{\beta}_1$, $\hat{\beta}_0$, s_ϵ^2, t, and r^2 did you observe?

Percentage of Type II errors:

(continued)

Discuss the differences between the results of this experiment and those of Experiment 8.7. What is the effect on increasing the spread of x? What does this suggest about ways to improve the analysis without increasing the sample size?